Undergraduate Texts in Physics

Series Editors

Kurt H. Becker, NYU Polytechnic School of Engineering, Brooklyn, NY, USA

Jean-Marc Di Meglio, Matière et Systèmes Complexes, Université Paris Diderot, Bâtiment Condorcet, Paris, France

Sadri D. Hassani, Department of Physics, Loomis Laboratory, University of Illinois at Urbana-Champaign, Urbana, IL, USA

Morten Hjorth-Jensen, Department of Physics, Blindern, University of Oslo, Oslo, Norway

Michael Inglis, Patchogue, NY, USA

Bill Munro, NTT Basic Research Laboratories, Optical Science Laboratories, Atsugi, Kanagawa, Japan

Susan Scott, Department of Quantum Science, Australian National University, Acton, ACT, Australia

Martin Stutzmann, Walter Schottky Institute, Technical University of Munich, Garching, Bayern, Germany

Undergraduate Texts in Physics (UTP) publishes authoritative texts covering topics encountered in a physics undergraduate syllabus. Each title in the series is suitable as an adopted text for undergraduate courses, typically containing practice problems, worked examples, chapter summaries, and suggestions for further reading. UTP titles should provide an exceptionally clear and concise treatment of a subject at undergraduate level, usually based on a successful lecture course. Core and elective subjects are considered for inclusion in UTP.

UTP books will be ideal candidates for course adoption, providing lecturers with a firm basis for development of lecture series, and students with an essential reference for their studies and beyond.

Francesco Scotognella

Exercises in Classical Physics—Mechanics and Thermodynamics

Basics of Theory and Solved Problems

 Springer

Francesco Scotognella
Department of Applied Science
and Technology
Politecnico di Torino
Torino, Italy

ISSN 2510-411X ISSN 2510-4128 (electronic)
Undergraduate Texts in Physics
ISBN 978-3-031-35076-4 ISBN 978-3-031-35074-0 (eBook)
https://doi.org/10.1007/978-3-031-35074-0

© The Editor(s) (if applicable) and The Author(s), under exclusive license to Springer Nature Switzerland AG 2023

This work is subject to copyright. All rights are solely and exclusively licensed by the Publisher, whether the whole or part of the material is concerned, specifically the rights of translation, reprinting, reuse of illustrations, recitation, broadcasting, reproduction on microfilms or in any other physical way, and transmission or information storage and retrieval, electronic adaptation, computer software, or by similar or dissimilar methodology now known or hereafter developed.
The use of general descriptive names, registered names, trademarks, service marks, etc. in this publication does not imply, even in the absence of a specific statement, that such names are exempt from the relevant protective laws and regulations and therefore free for general use.
The publisher, the authors, and the editors are safe to assume that the advice and information in this book are believed to be true and accurate at the date of publication. Neither the publisher nor the authors or the editors give a warranty, expressed or implied, with respect to the material contained herein or for any errors or omissions that may have been made. The publisher remains neutral with regard to jurisdictional claims in published maps and institutional affiliations.

This Springer imprint is published by the registered company Springer Nature Switzerland AG
The registered company address is: Gewerbestrasse 11, 6330 Cham, Switzerland

Paper in this product is recyclable.

Introduction

Physics aims to apply the scientific method to the observed phenomena of nature:

(i) experiment;
(ii) abstraction and formulation, in mathematical terms, of a law;
(iii) prediction of phenomena.

In mechanics, the motion of the bodies is studied. In an established treatment in most textbooks, mechanics is divided into kinematics and dynamics. Kinematics is a spatial-temporal description of the motion of bodies. Dynamics is the investigation of the causes that generate the motion of bodies. Fluid mechanics concerns with the mechanics of liquids and gases. Finally, thermodynamics studies the transformations, in terms of mass and energy, of a system.

The textbook, in its nine chapters, makes a synthesis of the above topics, dealing with material point kinematics and dynamics, introducing the concepts of work and energy, rediscussing the knowledge of kinematics and dynamics for point systems, and introducing a fundamental interaction between masses such as gravitational interaction. The motions of bodies will be described with respect to inertial and non-inertial reference systems (i.e., systems under acceleration with respect to inertial systems). Finally, the mechanics of fluids and gases will be discussed.

For each chapter, after the brief introduction, step-by-step examples and short exercises are offered, for which the solution is given.

Contents

1   **Kinematics of a Material Point** 1
2   **Dynamics of a Material Point** 15
3   **Work and Energy** 25
4   **Systems of Material Points** 35
5   **Gravity** 43
6   **Relative Motions** 55
7   **Rigid Body** 63
8   **Fluid Mechanics** 75
9   **Thermodynamics** 81

Appendix A 97
Appendix B 99
Appendix C 101
Appendix D 103
Appendix E 107
Index 109

Chapter 1
Kinematics of a Material Point

The simplest object that can be studied in mechanics is the material point. The actual size of the material point is neglected. The position of the material point (Fig. 1.1) is defined by the vector:

$$\vec{r} = \vec{r}_t = r_t \hat{u}_r = \begin{cases} x = x_t \\ y = y_t \\ z = z_t \end{cases} = r_x \hat{u}_x + r_y \hat{u}_y + r_z \hat{u}_z \quad (1.1)$$

$\hat{u}_r$ is the unit vector, i.e., vector with unit modulus:

$$\sqrt{u_x^2 + u_y^2 + u_z^2} = |\hat{u}_r| = u_r \quad (1.2)$$

$\hat{u}_r$ has the same direction and direction of $\vec{r}$. The unit of measurement of $\vec{r}$ is meter, [m].

From the position vector, we can define the average velocity:

$$\vec{v}_m = \frac{\vec{r}_{t\prime} - \vec{r}_t}{t\prime - t} = \frac{\vec{r}_{t+\Delta t} - \vec{r}_t}{t\prime - t} = \frac{\Delta \vec{r}}{\Delta t} \quad (1.3)$$

$\Delta \vec{r} = \vec{r}_{t\prime} - \vec{r}_t = \vec{r}_{t+\Delta t} - \vec{r}_t$ is the displacement.
The instantaneous velocity is

$$\vec{v} = \lim_{\Delta t \to 0} \frac{\vec{r}_{t\prime} - \vec{r}_t}{t\prime - t} = \lim_{\Delta t \to 0} \frac{\vec{r}_{t+\Delta t} - \vec{r}_t}{t\prime - t} = \lim_{\Delta t \to 0} \frac{\Delta \vec{r}}{\Delta t} = \frac{d\vec{r}}{dt} = \frac{ds}{dt} \hat{u}_T \quad (1.4)$$

s is the displacement. $\hat{u}_T$ is the unit vector related to the tangent direction of the trajectory. The unit of measurement of velocity is meter/second, [m]/[s].

Fig. 1.1 Reference frame *xyz* with the point P moving along the trajectory in red. In black the position vector $\vec{r}$ and in blue the velocity vector $\vec{v}$

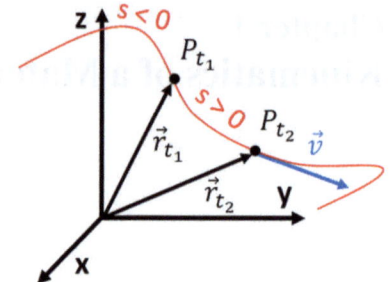

Fig. 1.2 (left) Velocity vector in a circular trajectory; (right) normal acceleration vector in a circumference with radius v (velocity)

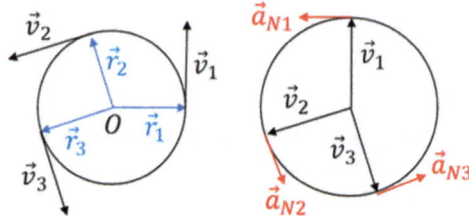

Analogously, the average acceleration is

$$\vec{a}_m = \frac{\Delta \vec{v}}{\Delta t} \qquad (1.5)$$

And the instantaneous acceleration:

$$\vec{a} = \frac{d\vec{v}}{dt} = \frac{d^2\vec{r}}{dt^2} = \frac{d}{dt}[v\hat{u}_{t,T}]$$

$$= \frac{dv}{dt}\hat{u}_{t,T} + v\frac{d\hat{u}_{t,T}}{dt} = \frac{dv}{dt}\hat{u}_T + \frac{v^2}{R}\hat{u}_N \qquad (1.6)$$

The unit of measurement of velocity is meter/second^2, [m]/[s]^2. $\hat{u}_N$ is the unit vector related to the normal direction of the trajectory.

Tangent acceleration: $\vec{a}_T = \frac{dv}{dt}\hat{u}_T$

Normal acceleration: $\vec{a}_N = \frac{v^2}{R}\hat{u}_N$.

Proof

The components of the acceleration can be derived in the simple case of velocity with constant modulus (Fig. 1.2).

The circumference on the left describes the trajectory. Its length is $2\pi R$ and the velocity (always tangent to the circumference) is given by

$$\frac{2\pi R}{T} = v \qquad (1.7)$$

1 Kinematics of a Material Point

T is the period: time interval for one complete revolution of circumference (in seconds).

For the circumference on the right, with radius v (velocity), normal acceleration, given by the change in direction of velocity, is always tangent. Similarly, it can be written as

$$\frac{2\pi v}{T} = a_N \tag{1.8}$$

Isolating T in the first equation

$$T = \frac{2\pi R}{v} \tag{1.9}$$

Thus, it is possible to find a_N as a function of v and R

$$a_N = \frac{2\pi v v}{2\pi R} = \frac{v^2}{R} \tag{1.10}$$

The modulus of the acceleration is

$$a = \sqrt{a_T^2 + a_N^2} = \sqrt{\left(\frac{dv}{dt}\right)^2 + \left(\frac{v^2}{R}\right)^2} \tag{1.11}$$

Kinematic using integration

From velocity to position; from acceleration to velocity:

$$\vec{r}_t = \vec{r}_{t_0} + \int_{t_0}^{t} \vec{v}_t \, dt \tag{1.12}$$

$$\vec{v}_t = \vec{v}_{t_0} + \int_{t_0}^{t} \vec{a}_t \, dt \tag{1.13}$$

$\vec{r}_{t_0}$ is the position vector at time $t = t_0$, while $\vec{v}_{t_0}$ is the velocity vector at time $t = t_0$.

With a constant velocity v_0, it is possible to define the linear motion: $s_t = s_0 + v_0 t$.

With a constant acceleration a_0, it is possible to define the uniformly accelerated linear motion:

$$s_t = s_0 + v_0 t + \frac{1}{2} a_0 t^2 \tag{1.14}$$

Proof
To derivate the formula of the uniformly accelerated linear motion:

$$\frac{d^2s}{dt^2} = \frac{dv}{dt} = a_0 \to dv = a_0 dt$$

$$\to \int_{t_0}^{t} dv = \int_{t_0}^{t} a_0 dt$$

$$\to v_t = v_{t_0} + a_0(t - t_0) \tag{1.15}$$

$$\frac{ds}{dt} = v_t = v_{t_0} + a_0(t - t_0)$$

$$\to ds = \left[v_{t_0} + a_0(t - t_0)\right] dt$$

$$\to \int_{t_0}^{t} ds = \int_{t_0}^{t} \left[v_{t_0} + a_0(t - t_0)\right] dt$$

$$\to s_t = s_{t_0} + v_0(t - t_0) + \frac{1}{2}a_0(t - t_0)^2 \tag{1.16}$$

$$if\ t_0 = 0 \Rightarrow s_t = s_0 + v_0 t + \frac{1}{2}a_0 t^2 \tag{1.17}$$

Motions on a plain

The parabolic motion (e.g., motion of a bullet) can be described in the following way:

$$\begin{cases} \vec{a} = -g\hat{u}_y \\ \vec{v} = v_x \hat{u}_x + v_y \hat{u}_y \end{cases} \tag{1.18}$$

where g is the gravitational acceleration: $g = 9.81 \frac{m}{s^2}$.

It is evident that the parabolic motion is the composition of a linear motion along x (where only the velocity v_x occurs) and a uniformly accelerated linear motion along y (where the velocity v_y and the acceleration g occur):

$$y = \frac{vx \operatorname{sen}\alpha}{v \cos\alpha} - \frac{1}{2}g\frac{x^2}{v^2 \cos^2\alpha} = x tg\alpha - \frac{1}{2}g\frac{x^2}{v^2 \cos^2\alpha} \tag{1.19}$$

Important points in the parabolic motion are:

– Projection on x of the maximum height:

$$x_V = -\frac{B}{2A} = \frac{tg\alpha \cdot 2 \cdot v^2 \cdot \cos^2\alpha}{2g}$$

$$= \frac{v^2}{2g} 2\cos\alpha \sin\alpha = \frac{v^2}{2g}\sin 2\alpha \tag{1.20}$$

1 Kinematics of a Material Point

Fig. 1.3 Position vector $\vec{r}_t$ and corresponding angular position ϑ_t in a circular motion

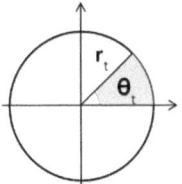

- Maximum height:

$$H = y(x_V) = \frac{v^2}{2g}\sin^2\alpha \qquad (1.21)$$

- Aerial phase length:

$$L = 2x_V = \frac{v^2}{g}\sin 2\alpha \qquad (1.22)$$

- Maeximum aerial phase length:

$$L_{\max} \text{ for } 2a = \pi/2, \text{ i.e., } a = \pi/4 = 45° \qquad (1.23)$$

Circular motion

On a circumference, the position of $\vec{r}_t$ can be described with the angle ϑ_t.
The angular velocity (direction perpendicular to the plane of motion) (Fig. 1.3) is:

$$\vec{\omega} = \frac{d\vartheta}{dt}\hat{u}_z \qquad (1.24)$$

With a unit of measurement [rad]/[s], while the angular acceleration is

$$\vec{\alpha} = \frac{d\omega}{dt}\hat{u}_z \qquad (1.25)$$

With a unit of measurement $[rad]/[s]^2$. The relations between velocity and angular velocity and between acceleration and angular acceleration are (Fig. 1.4).

$$\vec{v} = \vec{\omega} \times \vec{r} \qquad (1.26)$$

$$\vec{a} = \frac{d\vec{v}}{dt} = \frac{d}{dt}(\vec{\omega} \times \vec{r}) = \frac{d\vec{\omega}}{dt} \times \vec{r} + \vec{\omega} \times \frac{d\vec{r}}{dt}$$

Fig. 1.4 Position vector $\vec{r}$ (red dashed line), angular velocity vector $\vec{\omega}$ (green), and velocity vector $\vec{v}$ (purple solid line) in a circular motion. The three vectors follow the vector product described in Eq. 1.26

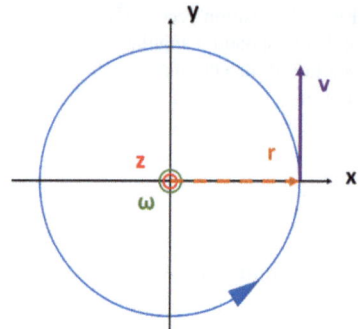

$$= \vec{\alpha} \times \vec{r} + \vec{\omega} \times \vec{v} = \vec{\alpha} \times \vec{r} + \vec{\omega} \times (\vec{\omega} \times \vec{r}) \qquad (1.27)$$

"×" relates to the vector product of two vectors.

The vector product can be calculated in the following way:

$$\vec{c} = \vec{a} \times \vec{b} = det \begin{bmatrix} \vec{u}_x & \vec{u}_y & \vec{u}_z \\ a_x & a_y & a_z \\ b_x & b_y & b_z \end{bmatrix}$$

$$= \vec{u}_x (a_y b_z - a_z b_y)$$
$$- \vec{u}_y (a_x b_z - a_z b_x)$$
$$+ \vec{u}_z (a_x b_y - a_y b_x) \qquad (1.28)$$

where $det[...]$ is the determinant of a matrix, although it should be stressed that the determinant of a matrix is a number. Here, the same method of calculation (for a 3×3 matrix) is employed, but the result is a vector. Thus, it is usually said that this method implies an abuse of notation: The method to calculate a determinant is used, but the result is not a determinant.

The vector product is also defined as

$$\vec{c} = \vec{a} \times \vec{b} = |\vec{a}||\vec{b}|\hat{n} sen\beta \qquad (1.29)$$

where β is the angle between $\vec{a}$ and $\vec{b}$, and $\hat{n}$ is the unit vector normal to the plane of $\vec{a}$ and $\vec{b}$.

Uniform circular motion:

The uniform circular motion is a motion with circular trajectory and constant angular velocity:

$$\vartheta_t = \vartheta_{t_0} + \omega(t - t_0) \qquad (1.30)$$

1 Kinematics of a Material Point

The uniform circular motion is a periodic motion: $\vec{r}_{t+nT} = \vec{r}_t$ (with $n \in \mathbb{N}$). The period (in seconds) is

$$T = \frac{2\pi}{\omega} \tag{1.31}$$

The frequency, reciprocal quantity to the period, is

$$\nu = \frac{1}{T} = \frac{\omega}{2\pi} \tag{1.32}$$

The unit is Hz (Hertz), which is 1/s.
Uniformly accelerated circular motion hourly law:

$$\vartheta_t = \vartheta_{t_0} + \omega(t - t_0) + \frac{1}{2}\alpha(t - t_0)^2 \tag{1.33}$$

Simple harmonic motion:

The simple harmonic motion is a motion around a point of equilibrium. It is a periodic motion with a sinusoidal (or co-sinusoidal) trend.

In simple harmonic motion, displacement, velocity, and acceleration can be written as

$$s_t = A\cos(\omega t + \phi) \tag{1.34}$$

$$v_t = \frac{ds_t}{dt} = -A\omega\sin(\omega t + \phi) \tag{1.35}$$

$$a_t = \frac{dv_t}{dt} = -A\omega^2\cos(\omega t + \phi) = -\omega^2 s_t \tag{1.36}$$

where A is the amplitude of the oscillation of the displacement.

From Eq. 1.35, it is clear that a linear proportionality occurs between the displacement s and the acceleration a. From this relation, the harmonic motion equation can be found:

$$\frac{d^2 s}{dt^2} = -\omega^2 s \tag{1.37}$$

From the plot of displacement s, velocity v, and acceleration a (please, see Fig. 1.5), it is evident the $\pi/2$ phase shift between s and v, the $\pi/2$ phase shift between v and a, and the π phase shift between s and a.

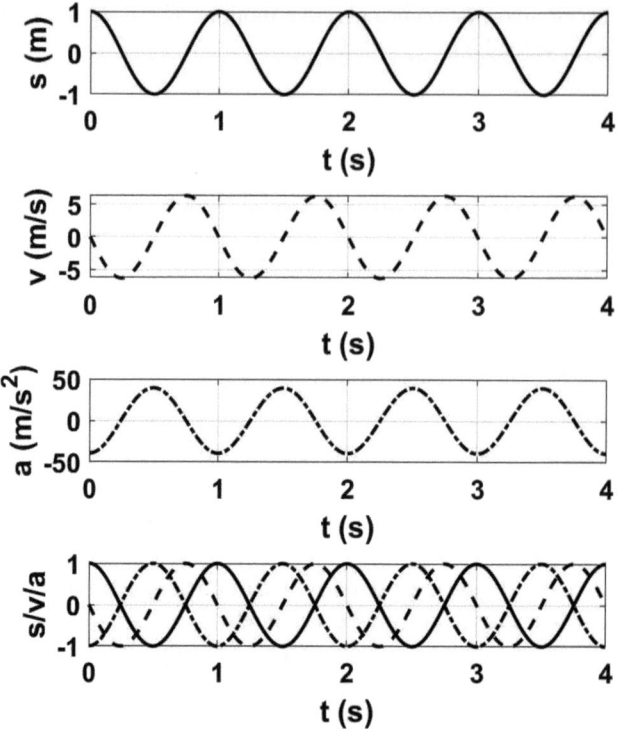

Fig. 1.5 Displacement s_t as described in Eq. 1.33; velocity v_t as described in Eq. 1.34; velocity a_t as described in Eq. 1.35. In the last plot at the bottom, the three functions are depicted to highlight the phase shift of $\pi/2$ between displacement and velocity and between velocity and acceleration

Examples

E1.1) The displacement of the body follows the function $s_t = 6t^4 + sint$. Determine velocity and acceleration.

Solution

Velocity: $v_t = 24t^3 + cost$
Acceleration: $a_t = 72t^2 - sint$.

E1.2) A car is traveling at a velocity $v = 100$ km/h. If the car is able to arrest in $s = 200$ m, calculate its deceleration (assuming a constant deceleration).

Solution

The two equations

$$s = v_0 t - \frac{1}{2}at^2$$

can be combined to obtain

$$v = v_0 - at$$

$$v^2 - v_0^2 = 2as$$

Thus

$$a = \frac{v^2 - v_0^2}{2s} = \frac{\left[0\frac{m}{s} - \left(27.78\frac{m}{s}\right)^2\right]}{2(200m)} = -1.93\frac{m}{s^2}$$

E1.3) In a circular trajectory, the position vector of a material point is $\vec{r} = 4\vec{u}_x + 5\vec{u}_y + 6\vec{u}_z$, and its angular velocity vector is $\vec{\omega} = 1\vec{u}_x + 2\vec{u}_y + 3\vec{u}_z$. Determine the vector of the velocity.

Solution

$$\vec{v} = \vec{\omega} \times \vec{r} = \det\begin{bmatrix} \vec{u}_x & \vec{u}_y & \vec{u}_z \\ 1 & 2 & 3 \\ 4 & 5 & 6 \end{bmatrix}$$

$$= \vec{u}_x(12 - 15) - \vec{u}_y(6 - 12) + \vec{u}_z(5 - 8)$$

$$= -3\vec{u}_x + 6\vec{u}_y - 3\vec{u}_z$$

E1.4) In a circular trajectory, the position vector of a material point is $\vec{r} = 5\vec{u}_y$, its angular velocity vector is $\vec{\omega} = 3\vec{u}_x$, and its angular acceleration vector is $\vec{\alpha} = 1\vec{u}_x$. Determine the modulus, direction, and verse of the vector of the acceleration.

Solution

$$\vec{a} = \vec{\alpha} \times \vec{r} + \vec{\omega} \times (\vec{\omega} \times \vec{r})$$

$$= \det\begin{bmatrix} \vec{u}_x & \vec{u}_y & \vec{u}_z \\ 1 & 0 & 0 \\ 0 & 5 & 0 \end{bmatrix} + \vec{\omega} \times \det\begin{bmatrix} \vec{u}_x & \vec{u}_y & \vec{u}_z \\ 3 & 0 & 0 \\ 0 & 5 & 0 \end{bmatrix}$$

$$= 5\vec{u}_z + \vec{\omega} \times 15\vec{u}_z = 5\vec{u}_z + \det\begin{bmatrix} \vec{u}_x & \vec{u}_y & \vec{u}_z \\ 3 & 0 & 0 \\ 0 & 0 & 15 \end{bmatrix}$$

$$= 5\vec{u}_z - 45\vec{u}_y$$

The modulus is $|\vec{a}| = \sqrt{5^2 + 45^2} = 45.28 \text{m/s}^2$.

E1.5) Derive the acceleration vector on an arbitrary trajectory.

Solution

The acceleration can be written as

$$\vec{a} = \frac{d\vec{v}}{dt}$$

where $\vec{v} = v\hat{u}_{t,T}$. In general, both v and $\hat{u}_{t,T}$ can depend on time: a change of v is related to a change of modulus and/or versus of the velocity, while a change of $\hat{u}_{t,T}$ is related to a change in the direction of the velocity. Thus, the acceleration can be rewritten as

$$\vec{a} = \frac{d\vec{v}}{dt} = \frac{d^2\vec{r}}{dt^2} = \frac{d}{dt}[v\hat{u}_{t,T}] = \frac{dv}{dt}\hat{u}_{t,T} + v\frac{d\hat{u}_{t,T}}{dt}$$

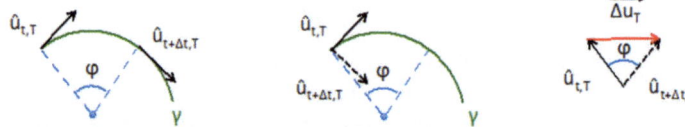

The difference between $\hat{u}_{t,T}$ and $\hat{u}_{t+\Delta t,T} - \hat{u}_{t,T}$ in a time interval Δt on the trajectory γ:

$$\Delta \vec{u}_T = \hat{u}_{t+\Delta t,T} - \hat{u}_{t,T}$$

For $\Delta t \to 0$, $\Delta \mathbf{u}_T$ tends to the infinitesimal vector $d\mathbf{u}_T$:

$$d\vec{u}_T = \hat{u}_{t+dt,T} - \hat{u}_{t,T}$$

$d\mathbf{u}_T$ is orthogonal to $\mathbf{u}_T$ (and parallel to $\mathbf{u}_N$, the versor that is orthogonal to the trajectory). Moreover, the modulus of $d\vec{u}_T$ can be approximated to

$$|d\vec{u}_T| \cong |\hat{u}_{t,T}|d\phi$$

Since $|\hat{u}_{t,T}| = 1$

$$|d\vec{u}_T| = d\phi$$

Consequently:

$$d\vec{u}_T = d\phi \hat{u}_N$$

1 Kinematics of a Material Point

And

$$\frac{d\hat{u}_T}{dt} = \frac{d\phi}{dt}\hat{u}_N$$

In this way, the acceleration can be rewritten as

$$\vec{a} = \frac{d\vec{v}}{dt} = \frac{d^2\vec{r}}{dt^2} = \frac{dv}{dt}\hat{u}_{t,T} + v\frac{d\phi}{dt}\hat{u}_N$$

$\frac{d\phi}{dt}$ can be rewritten in the following way:

$$\frac{d\phi}{dt} = \frac{d\phi}{ds}\frac{ds}{dt}$$

where $ds = d\varphi R$. Thus

$$\frac{d\phi}{ds}\frac{ds}{dt} = \frac{d\phi}{d\phi R}\frac{ds}{dt} = \frac{1}{R}v$$

in which R is the curvature radius of the trajectory.

Consequently, the acceleration is given by

$$\vec{a} = \frac{dv}{dt}\hat{u}_T + \frac{v^2}{R}\hat{u}_N$$

Important: in a general trajectory, the curvature radius R is not constant.

E1.6) In the harmonic motion of a material point along the x-axis, at $t_0 = 0$, the displacement of the material point is $x_{t0} = 0$. The pulsation of the motion is $\omega = 0.2$ rad/s, and the amplitude of the oscillation is $A = 1$ m. Find the acceleration at $t = 3$ s.

Solution

Since at t_0 the displacement is zero, the function x_t can be written as

$$x_t = A\sin(\omega t)$$

Thus, the acceleration can be written as

$$a_t = \frac{d^2 x_t}{dt^2} = -A\omega^2\sin(\omega t)$$

At $t = 3$ s

$$a_t = -A\omega^2 \sin(\omega t)$$

$$= -(1m)(0.2 \text{ rad/s})^2 \sin[(0.2 \text{ rad/s})(3 \text{ s})] = -0.02 \text{ m/s}^2$$

E1.7) A cannon fires a projectile with velocity

$$\vec{v} = (4\hat{u}_x + 20\hat{u}_y) \text{ m/s}$$

Find the coordinates of the bullet 2 s after it is fired.

Solution

Along x the velocity is $\vec{v}_x = (4\hat{u}_x)$ m/s. Thus

$$x_t = x_0 + v_x t = v_x t = (4 \text{ m/s})(2 \text{ s}) = 8 \text{ m}$$

Along y the velocity is $\vec{v}_y = (20\hat{u}_y)$ m/s, and the acceleration is $g = (-9.81\hat{u}_y)$ m/s^2. Thus

$$y_t = y_0 + v_y t - \frac{1}{2}gt^2 = v_y t - \frac{1}{2}gt^2$$

$$= (20 \text{ m/s})(2 \text{ s}) - \frac{1}{2}(9.81 \text{ m/s}^2)(2 \text{ s})^2 = 20.38 \text{ m}$$

Consequently, the coordinate of the projectile after 2 s is

$$\begin{cases} x = 8 \text{ m} \\ y = 20.38 \text{ m} \end{cases}$$

Exercises

(1.1) Determine the depth of a well knowing that a time $t = 4.8$ s elapses between the instant when a stone is dropped (with zero initial velocity) and the instant when the noise is heard, as a result of the stone hitting the bottom of the well (the speed of sound in air is 340 m/s and air resistance is negligible).

(A) 23.6 m
(B) 99.6 m
(C) 23,000 m
(D) 267 m
(E) 318 m

1 Kinematics of a Material Point

(1.2) Carl Lewis in 1991 set the 100 m record of 9.86 s. Usain Bolt in 2009 broke this record by doing the 100 m in 9.58 s. If they had competed at the same time, how many meters from the finish line would Lewis have been at the time Bolt crossed the finish line? Assume a constant speed.

(A) 1,44 m
(B) 2,84 m
(C) 3,98 m
(D) 4,23 m
(E) 28 m

(1.3) A marble A is dropped from a height h with zero initial velocity. At the same time, a second marble B is dropped from the ground upward with initial velocity v0. If the motion of the marbles occurs on the same vertical, determine the value of v0 so that the two marbles collide at half height (h/2).

(A) $\sqrt{gh}$
(B) $\frac{1}{2}\sqrt{gh}$
(C) $\sqrt{g/2h}$
(D) $3g$
(E) $4gh$

(1.4) Determine v_0 of exercise 3 that the two marbles collide when they have equal and opposite velocities.

(A) 0.5
(B) $\sqrt{gh}$
(C) $\sqrt{2gh}$
(D) $\frac{1}{2}\sqrt{gh}$
(E) $\sqrt{g/2h}$

(1.5) While riding a bicycle at a speed of 36 km/h, the yellow light goes off. Knowing that 5 s elapse before the red light and that you are 60 m from the traffic light, what is the minimum acceleration (in m/s^2) that would allow you to pass?

(A) 12.4
(B) 4.3
(C) 2.5
(D) 1.7
(E) 0.8

(1.6) What is the speed (in km/h) achieved in exercise 1.5?

(A) 20.3
(B) 33.1
(C) 42.7
(D) 50.4
(E) 61.8

(1.7) A material point travels along a curved trajectory. On a curved section with a radius of curvature equal to R = 1 m, the scalar velocity is $\sqrt{3}$ m/s. The acceleration vector forms an angle of about 37° with that of the velocity. What is the value of its acceleration in modulus?

(A) 100 m/s^2
(B) −4 m/s^2
(C) 5m/s^2
(D) 2 m/s^2
(E) 11.3 m/s^2

Chapter 2
Dynamics of a Material Point

The first law of dynamics

In an inertial reference frame, a free body moves in uniform rectilinear motion. If the speed of motion is zero, the body is at rest.

The momentum is:

$$\vec{q} = m\vec{v} \qquad (2.1)$$

Principle of conservation of momentum: The total momentum of the isolated system (where bodies are subjected to mutual interaction) is conserved.

The second principle of dynamics

It introduces the concept of **force** that expresses the interaction between bodies and accounts for the variation of momentum:

$$\vec{F} = \frac{d\vec{q}}{dt} \qquad (2.2)$$

The third principle of dynamics

For a two-body system:

$$\vec{F}_1 = -\vec{F}_2 \qquad (2.3)$$

The third principle is also known as the action-reaction principle.
$\vec{F}_1$ acts on body 1 by the effect of body 2; $\vec{F}_2$ acts on body 2 by the effect of body 1.

> Food for thought: There are fluids that do not follow the second principle: For weak forces they are strong, while they become tougher for stronger forces. They are called non-Newtonian fluids. An example is the corn starch in water.

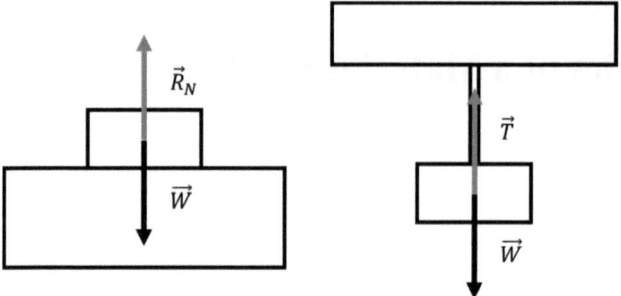

Fig. 2.1 (left) Equilibrium between the weight force $\vec{W}$ and the normal constraining rection $\vec{R}_N$ for a body resting on a table; (right) equilibrium between the weight force $\vec{W}$ and the tension $\vec{T}$ of an ideal rope (inextensible and of negligible mass) for a body hooked to the rope

Types of forces

Weight force

$$\vec{W} = m\vec{g} \qquad (2.4)$$

With the modulus of the gravitational acceleration, $g = 9.81 \frac{m}{s^2}$, such force is directed toward the center of the Earth, the largest gravitational attractor in the majority of the systems studied.

Constraining reactions

Such forces represent constraints, i.e., those forces whose effect on the motion of the body consists in limiting the motion to a certain region of space.

If a body is resting on a table, which in turn is resting on the Earth's ground with the upper face normal to gravitational acceleration, the weight force $\vec{W}$ due to the body is in equilibrium with the normal constraining reaction $\vec{R}_N$ due to the table (Fig. 2.1):

$$\vec{F}_{tot} = \vec{W} + \vec{R}_N = 0 \qquad (2.5)$$

s

The same can be said for the tension T of an ideal rope (inextensible and of negligible mass) to which a body of mass m is hooked (Fig. 2.1).

> Food for thought: What is the origin of the constraining reaction? A body on a table is not traveling toward the center of the Earth, but it is "blocked" by the table. This happens because the table, as other objects, is kept together by the chemical bonds, related to the electromagnetic force, which is much stronger than the gravitational force.

2 Dynamics of a Material Point

Sliding friction

The interface between two materials is not completely flat. The more the roughness, the more the friction.

Static sliding friction:

$$\left|\vec{R}_{T,max}\right| = \mu_s \left|\vec{R}_N\right| \text{ with } 0 < \mu_s < 1 \tag{2.6}$$

Dynamic sliding friction:

$$\left|\vec{R}_T\right| = \mu_d \left|\vec{R}_N\right| \text{ with } \mu_d < \mu_s \tag{2.7}$$

μ_s is the static sliding friction coefficient, while μ_d is the dynamic sliding friction coefficient. The two coefficients are effective and depend on the type of interface.

Viscous Friction (Stokes' Law)

The friction due to the motion in a medium can be described with Stokes' law:

$$\vec{F} = -\beta \vec{v} \tag{2.8}$$

β is a dissipative term (empirical, depends on the medium).

Elastic force

With a spring, or with an elastic material, the force depends linearly on the elongation/compression of the spring/material. Elastic force is described via Hooke's law (*ut tensio sic vis*: where there is tension there is a force):

$$\vec{F} = -k\vec{r} \tag{2.9}$$

k is the elastic constant, in N/m. k is a phenomenological value that depends on the material and the configuration of the spring.

With

$$\vec{r} = (l - l_0)\hat{u}_r \tag{2.10}$$

l_0 is the equilibrium length of the spring.

Springs in series:

$$\frac{1}{k_{eq}} = \sum_i \frac{1}{k_i} \tag{2.11}$$

Springs in parallel:

$$k_{eq} = \sum_i k_i \qquad (2.12)$$

With a body attached to a spring, constrained to the other side, harmonic motion can occur for the body left oscillating after elongating or compressing the spring:

$$m\frac{d^2x}{dt^2}\hat{u}_x = -kx\hat{u}_x = -m\omega^2 x_t \qquad (2.13)$$

(pulsation $\omega = \sqrt{k/m}$; period $T = 2\pi\sqrt{m/k}$).

Simple pendulum

A body with mass m is attached to a point by an inextensible and massless rope. With the second principle of dynamics, we can describe the forces:

$$m\vec{a} = \vec{T} - \vec{W} \qquad (2.14)$$

Thus, acceleration on a portion of the circumference occurs:

$$\vec{a} = \frac{d^2s}{dt^2}\hat{u}_t + \frac{1}{l}\left(\frac{ds}{dt}\right)^2 \hat{u}_n \qquad (2.15)$$

It is a harmonic motion:

$$l\frac{d^2\vartheta}{dt^2} = -g\vartheta \qquad (2.16)$$

with $\omega = \sqrt{g/l}$ and $T = 2\pi\sqrt{l/g}$.

With a spring free to oscillate, different types of harmonic motion can occur, depending on possible frictions and interactions that force the oscillation (Figs. 2.2, 2.3 and Table 2.1).

Examples

(**E2.1**) Two springs in parallel with elastic constants $k_1 = 1$ N/m and $k_2 = 3$ N/m, respectively, are in series with a third spring with elastic constant $k_3 = 4$ N/m. Calculate the equivalent elastic constant of the whole system.

Solution

Springs 1 and 2 in parallel:

$$k_{eq, parallel} = \sum_i k_i = k_1 + k_2 = 4 \, \text{N/m}$$

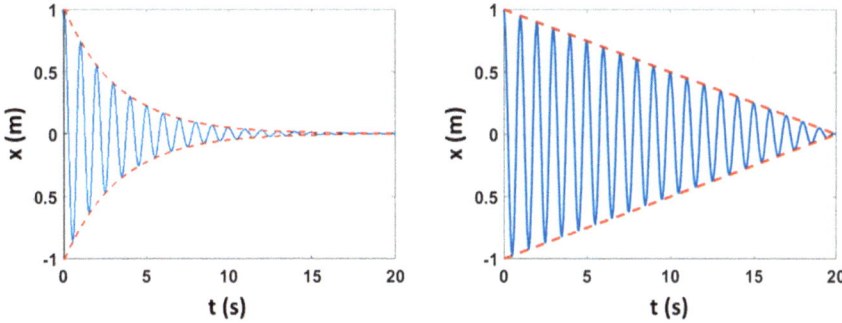

Fig. 2.2 A (left) Damped harmonic motion with viscous friction. (right) Damped harmonic motion with sliding friction

Fig. 2.3 Amplitude x as a function of the frequency ω for the forced-damped harmonic motion

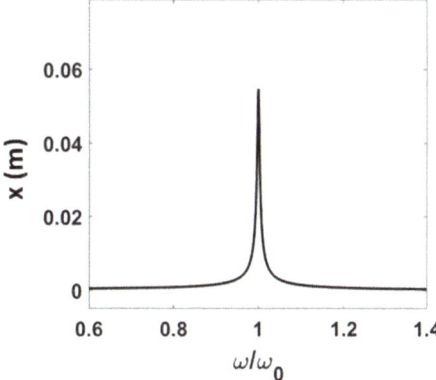

Springs 1 + 2 in series with spring 3:

$$\frac{1}{k_{eq}} = \sum_i \frac{1}{k_i}$$

$$k_{eq,series} = \frac{1}{\frac{1}{k_{eq,parallel}} + \frac{1}{k_3}} = 2\,\text{N/m}$$

(E2.2) A simple pendulum, with a body attached to a 5 m long rope, starts to oscillate with an oscillation amplitude of 1 m. Determine the maximum modulus of the velocity of the body in the oscillation.

Solution

In such a harmonic oscillator, the velocity is given by

$$v_t = \frac{ds_t}{dt} = -A\omega sen(\omega t + \phi)$$

Table 2.1 Different types of harmonic motions

Motion type	Motion equation	Features
Harmonic	$m\frac{d^2x}{dt^2}\hat{u}_x = -kx\hat{u}_x = -m\omega^2 x_t$	–Pulsation: $\omega_0 = \sqrt{k/m}$ –Period: $T = 2\pi\sqrt{m/k}$
Damped harmonic (with viscous friction, Fig. 2.2, left)	$\frac{d^2x}{dt^2} + \frac{\beta}{m}\frac{dx}{dt} + \frac{k}{m}x = 0$ $\Delta = \frac{\beta^2}{m^2} - 4\frac{k}{m}$	Three cases: –$\Delta < 0$: underdamped, oscillation with exponential damping (exponential envelope function of oscillation) –$\Delta = 0$: critical damping, returns to equilibrium as fast as possible without oscillating –$\Delta > 0$: overdamped, returns to balance without oscillating
Damped harmonic (with sliding friction, Fig. 2.2, right)	$\frac{d^2x}{dt^2} + \frac{k}{m}x + \frac{\mu_D R_N}{m} = 0$	Oscillation with linear damping
Forced harmonic	$\frac{d^2x}{dt^2} + \frac{k}{m}x = \frac{F}{m}$	–$F = F_0\cos(\omega t)$ forcing (in general $\omega \neq \omega_0$) –Oscillation amplitude $x_0 = \frac{F_0}{m(\omega_0^2 - \omega^2)}$ –For $\omega \to \omega_0$: resonance that leads to rupture $x \to \infty$
Forced-damped harmonic	$\frac{d^2x}{dt^2} + \frac{\beta}{m}\frac{dx}{dt} + \frac{k}{m}x = \frac{F}{m}$	–$F = F_0\cos(\omega t)$ forcing –For we have $\omega \to \omega_0 x_{max}$ [as in Fig. 2.3, peak resonance of function $x(\omega)$]

In a simple pendulum

$$\omega = \sqrt{g/l}$$

Thus

$$v_t = -A\sqrt{g/l}\sin\left(\sqrt{g/l}\,t + \phi\right)$$

In this case the initial phase ϕ is zero. Moreover, the maximum of the velocity function relates to the $sin(\sqrt{g/l}t) = 1$, which means angle 0, i.e., the body at the equilibrium position.

Thus

$$v_t = A\sqrt{g/l} = 1m\sqrt{\frac{9.8\,\text{m/s}^2}{5\,\text{m}}} = 1.4\,\text{m/s}$$

2 Dynamics of a Material Point

(E2.3) A block of m = 5 kg is resting on an inclined plane with angle a. There is friction between the block and the interface, and the coefficient of static sliding friction is $\mu_s = 0.2$. What is the minimum angle for the block to slide?

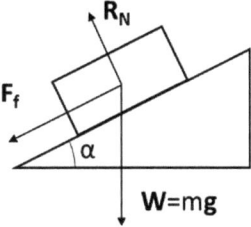

Solution

$$\left|\vec{F}_f\right| = mg\sin\alpha$$

$$\left|\vec{R}_N\right| = mg\cos\alpha$$

For the friction at the interface

$$\left|\vec{F}_f\right| = \mu_s \left|\vec{R}_N\right|$$

$\left|\vec{F}_f\right|/\left|\vec{R}_N\right| = \tan\alpha$, thus

$$\mu_s = \tan\alpha$$

And consequently

$$\alpha = \tan^{-1}\mu_s = 11.31°$$

(E2.4) Two masses m_1 and m_2 hang from the two ends of a wire, inextensible and of negligible mass, passing through a pulley. If the ratio of its masses is $m_2/m_1 = 3$, find the acceleration with which the two masses move.

Solution

$$\begin{cases} T - m_1 g = m_1 a \\ m_2 g - T = m_2 a \end{cases}$$

With T

$$\begin{cases} T = m_1 a + m_1 g \\ T = m_2 g - m_2 a \end{cases}$$

$$m_1 a + m_1 g = m_2 g - m_2 a$$

$$m_1 a + m_2 a = m_2 g - m_1 g$$

$$(m_1 + m_2) a = (m_2 - m_1) g$$

Thus

$$a = \frac{m_2 - m_1}{m_1 + m_2} g = \frac{\frac{m_2}{m_1} - 1}{\frac{m_2}{m_1} + 1} g$$

$$= \frac{3 - 1}{3 + 1} g = \frac{1}{2} \left(9.8 \, \frac{m}{s^2} \right) = 4.9 \, \frac{m}{s^2}$$

Exercises

(2.1) A watchmaker wants to adjust a pendulum clock that accumulates a delay of one minute every hour. The pendulum consists of a mass m hanging from a wire of length L. What adjustment can the watchmaker make?

(A) Shorten the thread by 3.3%
(B) Lengthen the thread by 5.2%
(C) Shorten the thread by 4.8%
(D) Lengthen the wire by 6.7%
(E) Lighten the mass by 8.2%

(2.2) In harmonic motion, the second derivative, with respect to time, of displacement is.

(A) as its velocity
(B) as the displacement multiplied by ω^2
(C) as the displacement multiplied by -1

(D) as the displacement multiplied by ω
(E) as its velocity multiplied by ω

(2.3) For the work of the sliding friction force, which of the following statements is false?

(A) it doubles if the trajectory traveled is doubled
(B) it is negative
(C) its calculation shows that the frictional force is not conservative
(D) it is independent of the scalar product between force and displacement
(E) it is directly proportional to the modulus of the constraining reaction

(2.4) What is the unit of measurement of the elastic constant k of a spring (remember Hooke's law)?

(A) N*m
(B) m^2/N
(C) J/s
(D) N
(E) Ns^2/ms^4

(2.5) A marble is launched down a ramp of length L = 2 m inclined at an angle $\alpha = 45°$ to the horizontal with velocity v0 = 15 m/s. Knowing that the dynamic friction coefficient between the ramp and the marble is worth $\mu d = 0.2$ for the first meter of the ramp and $\mu d = 0.4$ for the second meter, the speed of the marble at the top of the ramp is.

(A) 11.445 m/s
(B) 4.765 m/s
(C) 13.745 m/s
(D) 9.375 m/s
(E) 6.035 m/s

(2.6) All springs in this scheme have k = 1 N/m. The equivalent spring constant k_{eq} is.

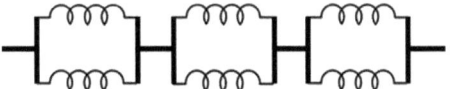

(A) 1 N/m
(B) 3 N/m
(C) 6 N/m
(D) ½ N/m
(E) 2/3 N/m

(2.7) What is the pulsation of the system in the figure?
($k_1 = 100$ N/m; $k_2 = 150$ N/m; $k_3 = 80$ N/m; $M = 12500$ g).

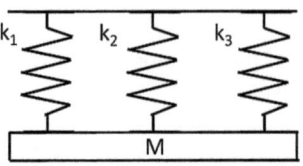

(A) 26.40 rad/s
(B) 2.74 rad/s
(C) 1.66 rad/s
(D) 5.14 rad/s
(E) 10.43 rad/s

Chapter 3
Work and Energy

A body is moved under the effect of a force. The elementary work is defined as

$$\delta W = \vec{F} \cdot d\vec{r} \tag{3.1}$$

· which relates to the scalar product between two vectors:

$$\delta W = |\vec{F}||d\vec{r}|\cos\vartheta \tag{3.2}$$

ϑ is the angle between the two vectors.

The integral work is given by the integral between the force and the displacement. In general, the work depends on the trajectory of the displacement γ

$$W_{A\gamma B} = \int_{A\gamma B} \vec{F} \cdot d\vec{r} \tag{3.3}$$

The power is given by

$$P = \frac{\delta W}{dt} = \vec{F} \cdot \frac{d\vec{r}}{dt} \tag{3.4}$$

The kinetic energy is a positive scalar quantity given by

$$E_k = \frac{1}{2}mv^2 \tag{3.5}$$

© The Author(s), under exclusive license to Springer Nature Switzerland AG 2023
F. Scotognella, *Exercises in Classical Physics—Mechanics and Thermodynamics*,
Undergraduate Texts in Physics, https://doi.org/10.1007/978-3-031-35074-0_3

Living forces theorem (relation between work and kinetic energy):

$$L_{A\gamma B} = \int_{A\gamma B} \delta L = \int_{A\gamma B} dE_k = E_{k,B} - E_{k,A} \qquad (3.6)$$

Proof

The living forces theorem can be easily demonstrated by introducing the projection of the applied force in the direction of the displacement F_T (T relates to the direction tangent to the displacement):

$$\delta L = F_T dr = m a_T dr = m \frac{dv}{dt} dr$$
$$= m \frac{dr}{dt} dv = m v dv \qquad (3.7)$$

$$L_{A\gamma B} = \int_{A\gamma B} m v \, dv = \frac{1}{2} m v_B^2 - \frac{1}{2} m v_A^2$$
$$= E_{k,B} - E_{k,A} = \Delta E_k \qquad (3.8)$$

If the calculated work does not depend on the trajectory, but only on the initial and final positions, the force is *conservative*.

Potential energy for conservative forces:

$$\delta L = -dE_p(\vec{r}) \qquad (3.9)$$

Weight force:

Force directed along $-z$:
$\mathbf{F} = (0, 0, -mg)$ (Fig. 3.1)

$$L_{A\gamma B} = \int_{A\gamma B} \vec{F} \cdot d\vec{r} = \int_{A\gamma B} m\vec{g} \cdot d\vec{r} \qquad (3.10)$$

Considering that the weight force has nonzero component only along $-z$, and dr (for the monotonic trajectory in the figure) has projection along the z-axis (positive direction, hence $+z$), therefore, the scalar product in the integral can be rewritten:

$$L_{A\gamma B} = -\int_{A\gamma B} mg \, dz = -mg \int_{z_A}^{z_B} dz$$

3 Work and Energy

Fig. 3.1 Weight and displacement $d\vec{r}$ in an arbitrary trajectory between point A and point B

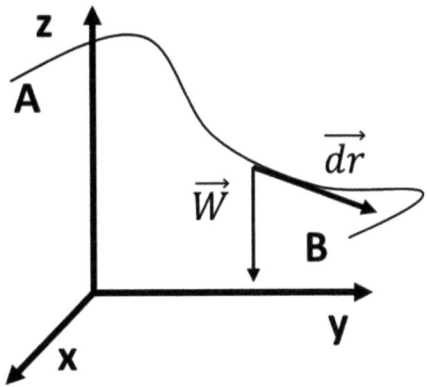

$$= -mg(z_B - z_A) = mg(z_A - z_B) \qquad (3.11)$$

$$L_{AB} = -\Delta E_p \Rightarrow \Delta E_p = mg(z_B - z_A) \qquad (3.12)$$

If $z_A = 0$ (ground level),

$$E_p = mgz \qquad (3.13)$$

Elastic force:

$\mathbf{F} = (-\mathbf{k}\mathbf{x}, 0, 0)$.
Force along $-x$:

$$L_{AyB} = -\int_{AyB} kx\,dx = -k\int_{x_A}^{x_B} x\,dx = -k\left[\frac{x^2}{2}\right]_{x_A}^{x_B}$$

$$= -\left(\frac{1}{2}kx_B^2 - \frac{1}{2}kx_A^2\right) = \frac{1}{2}kx_A^2 - \frac{1}{2}kx_B^2 \qquad (3.14)$$

$$L_{AB} = -\Delta E_p \Rightarrow \Delta E_p = \frac{1}{2}k(x_B^2 - x_A^2) \qquad (3.15)$$

Setting to zero the energy for the spring at equilibrium length:

$$E_p = \frac{1}{2}kx^2 \qquad (3.16)$$

Sliding friction:

$$L_{A\gamma B} = \int_{A\gamma B} \vec{F} \cdot d\vec{r} = -\int_{A\gamma B} \mu_d R_N \hat{u}_v \cdot d\vec{r}$$

$$= -\mu_d R_N \int_{A\gamma B} \hat{u}_v \cdot d\vec{r} \qquad (3.17)$$

(with constraining reaction R_N from A to B).

u$_v$ *depends on the trajectory.*

The displacement is

$$d\vec{r} = \vec{v}\, dt = v\hat{u}_v dt \qquad (3.18)$$

Thus, Eq. 3.17 becomes

$$\rightarrow L_{A\gamma B} = -\mu_d R_N \int_{A\gamma B} \hat{u}_v v \cdot \hat{u}_v dt \qquad (3.19)$$

in which $(\hat{u}_v \cdot \hat{u}_v) = 1$. Finally,

$$L_{A\gamma B} = -\mu_d R_N \int_{A\gamma B} v\, dt = -\mu_d R_N \ell_{A\gamma B} \qquad (3.20)$$

where $\ell_{A\gamma B}$ is the length of the trajectory. Different trajectories can lead to different $\ell_{A\gamma B}$ and, consequently, to different values of the work of the sliding friction. Thus, the work of the friction depends on the trajectory chosen to reach B starting from A and, consequently, the sliding friction is a non-conservative force.

For a conservative force on any closed trajectory (e.g., from A to B, then from B to A):

$$L_{closed\,trajectory} = L_{AB} + L_{BA}$$
$$= \int_A^B \vec{F} \cdot d\vec{r} + \int_B^A \vec{F} \cdot d\vec{r}$$
$$= \oint \vec{F} \cdot d\vec{r} = 0 \qquad (3.21)$$

since

$$\int_A^B \vec{F} \cdot d\vec{r} = -\int_B^A \vec{F} \cdot d\vec{r} \qquad (3.22)$$

3 Work and Energy

The total work done by a conservative force, when the path is a closed loop, is zero.

Food for thought: It is possible to determine the force from the potential energy using the derivation, i.e., the spatial variation of the potential energy:

$$\vec{F} = -\frac{dE_p}{dr}\vec{u}_F \tag{3.23}$$

The correct mathematical operation is the gradient:

$$\vec{F} = -grad E_p = -\vec{\nabla} E_p$$
$$= -\left(\frac{\partial E_p}{\partial x}\vec{u}_x + \frac{\partial E_p}{\partial y}\vec{u}_y + \frac{\partial E_p}{\partial z}\vec{u}_z\right) \tag{3.24}$$

Principle of conservation of mechanical energy

If only conservative forces act on a body: $dE = 0$. If both conservative and non-conservative forces act on a body: $dE = \delta L^{NC}$.

Mechanical energy of harmonic motion

The mechanical energy of harmonic motion can be determined knowing that there is only a conservative force (i.e., elastic force) and, thus, there is a conservation of energy

$$E = E_k + E_p = \frac{1}{2}mv^2 + \frac{1}{2}kx^2 = constant \tag{3.25}$$

Thus

$$E = \frac{1}{2}m[-A\omega sin(\omega t + \phi)]^2$$
$$+ \frac{1}{2}k[A cos(\omega t + \phi)]^2$$
$$= \frac{1}{2}mA^2\omega^2 sin^2(\omega t + \phi)$$
$$+ \frac{1}{2}kA^2 cos^2(\omega t + \phi) \tag{3.26}$$

Knowing that $\omega^2 = \frac{k}{m} \rightarrow m = \frac{k}{\omega^2}$

$$E = \frac{1}{2}\frac{k}{\omega^2}A^2\omega^2 sin^2(\omega t + \phi)$$
$$+ \frac{1}{2}kA^2 cos^2(\omega t + \phi)$$

$$= \frac{1}{2}kA^2\left[\cos^2(\omega t + \phi) + \sin^2(\omega t + \phi)\right] \quad (3.27)$$

$\cos^2(\omega t + \phi) + \sin^2(\omega t + \phi) = 1$. Thus

$$E = \frac{1}{2}kA^2 \quad (3.28)$$

Examples

E3.1) A body of 2 kg, under the effect of a force, accelerates with a = 3 m/s². The angle between the acceleration and the trajectory of the body is constantly π/3. If the displacement of the body is s = 5 m, how much work is produced by the body?

Solution

The work is

$$L_{AB} = \int_A^B \vec{F} \cdot d\vec{r} = m\,a\,s\,\cos(\pi/3)$$

$$= (2\,\text{kg})(3\,\text{m/s}^2)(5\,\text{m})0.5 = 15\,J$$

E3.2) A ball with a weight of 100 g is thrown upward by accompanying it 40 cm with the hand. When the ball leaves the hand, its velocity is 10 m/s. Find the force exerted by the hand and the height reached by the ball (setting the zero to the point where the hand and ball start to move).

Solution

Setting the $z = 0$ m where the hand and ball start to move, we have in this scenario

$$E_1 = E_{k,1} + E_{p,1} = 0 + 0$$

When the ball leaves the hand:

$$E_2 = E_{k,2} + E_{p,2}$$

$$= (0.5)(0.1\,\text{kg})(100\,\text{m}^2/\text{s}^2) + (0.1\,\text{kg})(9.8\,\text{m/s}^2)(0.4\text{m})$$

$$= 5.39\,J$$

This difference in energy $E_2 - E_1$ is equal to the work done by the hand along the displacement $\Delta h = 0.4$ m:

3 Work and Energy

$$W = E_2 - E_1 = E_2 = F\Delta h$$

Thus

$$F = E_2/\Delta h = 5.39 \, J/0.4 \, m = 13.47 \, N$$

The maximum is reached by the ball when all the kinetic energy $E_{k,2}$ is converted into gravitational potential energy:

$$E_{k,2} = (0.5)(0.1 \, \text{kg})(100 \, \text{m}^2/\text{s}^2) = 0.5 \, mv_2^2 = E_{p,3} = mg\Delta h'$$

Thus

$$\Delta h' = \frac{v_2^2}{2g} = \frac{100 \, \text{m}^2/\text{s}^2}{2(9.8 \, \text{m/s}^2)} = 5.1 \, \text{m}$$

The total height is

$$\Delta h_{max} = \Delta h + \Delta h' = 5.1 \, \text{m} + 0.4 \, \text{m} = 5.5 \, \text{m}$$

E3.3) Two springs in parallel, with spring constant $k_1 = 1$ N/m and $k_2 = 3$ N/m, respectively, swing a 1 kg body. If at a certain instant the body, at a distance x = 1 m from the equilibrium position, moves with a velocity of 1 m/s. Determine the maximum amplitude of oscillation of the body.

Solution

The equivalent elastic constant for the systems of springs is

$$k_{eq} = k_1 + k_2 = 4 \, \text{N/m}$$

Consequently:

$$\frac{1}{2}k_{eq}A^2 = E_p + E_k = \frac{1}{2}k_{eq}x^2 + \frac{1}{2}mv^2$$

$$= \frac{1}{2}(4 \, \text{N/m})(1 \, \text{m})^2 + \frac{1}{2}(1 \, \text{kg})(1 \, \text{m/s})^2$$

$$= 2 \, J + 0.5 \, J = 2.5 \, J$$

Thus

$$A = \sqrt{\frac{2(E_p + E_k)}{k_{eq}}} = \sqrt{\frac{2(2.5 \, J)}{4 \, \text{N/m}}} = 1.06 \, \text{m}$$

E3.4) A 70 kg American football player runs with the ball at a speed of 5 m/s. He is stopped in 1 s during a tackle. How much is the average power developed during the tackle?

Solution

In the tackle, a deceleration occurs:

$$s = vt - \frac{1}{2}at^2$$

Thus, the power is

$$P = \frac{L}{t} = \frac{Fs}{t} = \frac{Ft\left(v - \frac{1}{2}at\right)}{t} = F\left(v - \frac{1}{2}at\right)$$

The force is the momentum over time $F = \frac{mv}{t}$ (it is an average force). Thus

$$P = \frac{mv}{t}t\left(\frac{v}{t} - \frac{1}{2}a\right) = mv\left(\frac{v}{t} - \frac{1}{2}vt\right)$$

$$= (70\,\text{kg})\left(5\,\frac{\text{m}}{\text{s}}\right)\left[\frac{5\,\frac{\text{m}}{\text{s}}}{1\,\text{s}} - \frac{1}{2}\left(5\,\frac{\text{m}}{\text{s}}\right)(1\,\text{s})\right] = 875\,\text{W}$$

E3.5) The potential energy is described as a function that includes the following points:

Position x (m)	Potential energy (J)
−5	25
−3	9
1	1
4	16
6	36

Find the force that is related to such potential energy.

Solution

It is evident that the potential energy is described by a parabolic function

$$E_p = Ax^2$$

with $A = 1$ N.

The force can be found from the potential according to the relation

$$\vec{F} = -\frac{dE_p}{dr}\vec{u}_F$$

Thus

$$\vec{F} = -\frac{dE_p}{dx}\vec{u}_x = -2x\,\vec{u}_x$$

Exercises

(3.1) A mass of 500 g compresses by 5 cm a spring of elastic constant 200 N/m attached to the lower end of a plane inclined 60°. If the coefficient of kinetic friction between the plane and the mass is 0.3, determine the maximum height reached by the mass once it is left free.

(A) 12 cm
(B) 0.02 m
(C) 0.6 m
(D) 4 cm
(E) 22 cm

(3.2) An object of mass m = 4 kg is thrown along an upward-inclined plane. The inclined plane is 2 m high and forms an angle with the ground of 30°. The initial speed of the object is 15 m/s. How long does it take to reach the top of the inclined plane?

(A) 0.11 s
(B) 0.28 s
(C) 0.39 s
(D) 0.54 s
(E) 0.68 s

(3.3) A marble is launched down a ramp of length L = 3 m inclined at an angle α = 45° to the horizontal with velocity v_0= 20 m/s. Knowing that the dynamic friction coefficient between the ramp and the marble is worth μ_d= 0.3, determine the velocity in m/s with which the marble reaches the bottom of the ramp.

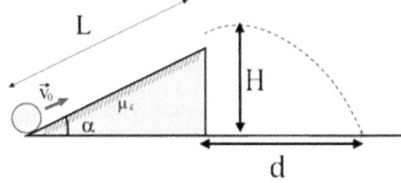

(A) 12.50
(B) 18.60
(C) 23.80
(D) 27.20
(E) 32.10

(3.4) Calculate the distance from ramp d where the marble falls to the ground (in m) in problem 3.3.

(A) 15.43
(B) 21.98
(C) 37.27
(D) 42.09
(E) 51.35

(3.5) A 70 kg skater is at the top of a U-shaped ramp at a height of 10 m. If the bottom of the ramp is at a height of 0.5 m, what is the maximum velocity reached by the skater?

(A) 13.65 m/s
(B) 25.63 m/s
(C) 30.04 m/s
(D) 37.12 m/s
(E) 41.96 m/s

(3.6) Which of these statements is false for power?

(A) it is the derivative of work with respect to time
(B) its unit of measurement is the watt
(C) it is the scalar product of force times velocity
(D) it is the scalar product of force times space
(E) its unit of measurement is Joule/second

Chapter 4
Systems of Material Points

The second principle of dynamics for systems of ith material points is given by

$$\vec{F}_i = m_i \vec{a}_i = m_i \frac{d^2 \vec{r}_i}{dt^2} \tag{4.1}$$

which relates to N second-order vectorial differential equations (=3N scalar equations):

$$\begin{cases} \vec{F}_{x,i} = m_i \vec{a}_{x,i} = m_i \frac{d^2 x_i}{dt^2} \hat{u}_x \\ \vec{F}_{y,i} = m_i \vec{a}_{y,i} = m_i \frac{d^2 y_i}{dt^2} \hat{u}_y \\ \vec{F}_{z,i} = m_i \vec{a}_{z,i} = m_i \frac{d^2 z_i}{dt^2} \hat{u}_z \end{cases} \tag{4.2}$$

The description of this system can be very complex. For many material points (≥ 3), the system of scalar equations can be solved numerically. A particular system is the rigid body, which is a system of material points in which the mutual distance between any two points remains constant during motion (Chap. 7).

External and internal forces should be defined:

$$\vec{F}_i = \vec{F}_i^{(E)} + \vec{F}_i^{(I)} \tag{4.3}$$

The internal forces relate to the interaction between the bodies of the system. Due to the third principle of dynamics,

$$\vec{F}^{(I)} = \sum_{\substack{i,j=1 \\ i \neq j}}^{N} \vec{f}_{ij} = 0 \tag{4.4}$$

Thus, only external forces remain.
Consequently, Euler's first law is written as

$$\vec{F}^{(E)} = \frac{d\vec{q}}{dt} \qquad (4.5)$$

The system of material points keeps being very complex to be studied. A useful tool is the introduction of a point that takes into account the whole system. Such a point is the center of mass.

The center of mass is defined as

$$\vec{r}_{CM} = \frac{\sum_{i=1}^{N} m_i \vec{r}_i}{\sum_{i=1}^{N} m_i} \qquad (4.6)$$

Thus, it is possible to introduce the center of mass theorem:

$$\vec{F}^{(E)} = \frac{d}{dt} M \vec{v}_{CM} = M \vec{a}_{CM} \qquad (4.7)$$

4.1 Collisions

For a system where external forces are not applied,

$$\vec{F}^{(E)} = \frac{d\vec{q}}{dt} = 0 \rightarrow \vec{q} = \text{constant} \qquad (4.8)$$

In such a system, collisions can occur between material points. It is possible to distinguish between elastic collisions and inelastic collisions. In the elastic collision, momentum and kinetic energy (of the system) are conserved:

$$\begin{cases} m_1 \vec{v}_1^{in} + m_2 \vec{v}_2^{in} = m_1 \vec{v}_1^{fin} + m_2 \vec{v}_2^{fin} \\ \frac{1}{2} m_1 v_1^{in^2} + \frac{1}{2} m_2 v_2^{in^2} = \frac{1}{2} m_1 v_1^{fin^2} + \frac{1}{2} m_2 v_2^{fin^2} \end{cases} \qquad (4.9)$$

In a head-on collision, the vector equation for the momentum (in Eq. 4.9) becomes a scalar equation (Fig. 4.1):

$$\begin{cases} m_1 v_1^{in} + m_2 v_2^{in} = m_1 v_1^{fin} + m_2 v_2^{fin} \\ \frac{1}{2} m_1 v_1^{in^2} + \frac{1}{2} m_2 v_2^{in^2} = \frac{1}{2} m_1 v_1^{fin^2} + \frac{1}{2} m_2 v_2^{fin^2} \end{cases} \qquad (4.10)$$

In the case of inelastic collision, the kinetic energy is not conserved. Part of the energy is converted in heat, sound, and deformations of the colliding bodies.

Fig. 4.1 Sketch of a head-on collision

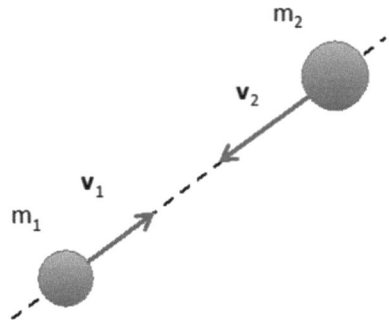

A particular case is the perfectly inelastic collision, in which two colliding bodies become a single body. Thus

$$v_1^{fin} = v_2^{fin} = v^{fin} \tag{4.11}$$

With the momentum conservation:

$$m_1 v_1^{in} + m_2 v_2^{in} = (m_1 + m_2) v^{fin} \tag{4.12}$$

Impulse theorem: the second law of dynamics implies that the impulse of a force applied to a material point causes the change in momentum:

$$d\vec{q} = \vec{F} dt \tag{4.13}$$

If $\vec{F}_{AB}$ is the force acting on the body B because of the interaction with the body A, the impulse related to this force is given by

$$\vec{I}_A = \int_{t_0}^{t_0+\tau} \vec{F}_{AB} dt = m_A \vec{v}_A^{fin} - m_A \vec{v}_A^{in} \tag{4.14}$$

Analogously, if $\vec{F}_{BA}$ is the force acting on the body A because of the interaction with the body B, the impulse related to this force is given by

$$\vec{I}_B = \int_{t_0}^{t_0+\tau} \vec{F}_{BA} dt = m_B \vec{v}_B^{fin} - m_B \vec{v}_B^{in} \tag{4.15}$$

Since $\vec{F}_{AB} = -\vec{F}_{BA}$, $\vec{I}_A = -\vec{I}_B$.

For each of the two bodies, there is a change in momentum $\Delta \vec{q}$ (the momentum of the system is conserved). If we assume that the collision occurs instantaneously, per $\tau \to 0$, $\mathbf{F} \to \infty$ (where τ is the collision duration).

In reality, the collision occurs in a time interval small but finite. Thus, the value of the force will then be very large, but finite. Forces that take on very large values for short time intervals are called impulsive forces.

Examples

(E4.1) A toy car of mass $m_1 = 50$ g travels on a magnetic track at speed $v_0 = 2$ m/s. At time $t = 0$ s, a switch "turns on" on the magnetic track a force that is impressed on the toy car. The force has the trend shown in the graph in the figure.

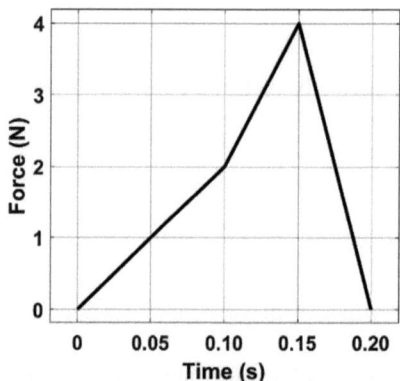

What is the speed of the little car after the force is turned off?

Solution

The impulse of the toy car

$$I = \int_{t=0s}^{t=0.2s} F\,dt = (7N)(0.2s) = 1.4\,Ns$$

Is equal to the momentum variation:

$$I = m_1 v_1 - m_1 v_0 = 0.05, kg \cdot v_1 - 0.05\,kg \cdot 2\frac{m}{s} = 1.4\,Ns$$

Thus $v_1 = 26\frac{m}{s}$.

(E4.2) The toy car in Example E4.1 bumps into and hooks onto another toy car of mass $m_2 = 100$ g. What is the loss of kinetic energy?

Solution

The momentum conservation is

4.1 Collisions

$$m_1 v_1 = (m_1 + m_2) v^{fin}$$

So

$$v^{fin} = \frac{m_1 v_1}{m_1 + m_2}$$

Thus $v^{fin} = 8.67 \frac{m}{s}$.
The loss of kinetic energy is

$$\Delta E = \frac{1}{2}(m_1 + m_2) v^{fin^2} - \frac{1}{2} m_1 v_1^2 = -28.16 \, J$$

(E4.3) The mass of an automobile is 1500 kg. If the automobile has a constant speed of 20 m/s, what is the impulse required to make its velocity reach 5 m/s?

Solution

The module of the initial momentum $\vec{q}_i$ of the automobile is given by

$$|\vec{q}_i| = mv = (1500 \, \text{kg})(20 \, \text{m/s}) = 3 \times 10^4 \frac{\text{kg m}}{\text{s}}$$

The module of the momentum $|\vec{q}_f|$ when the velocity of the automobile is 5 m/s is

$$|\vec{q}_f| = mv = (1500 \, \text{kg})(5 \, \text{m/s}) = 0.75 \times 10^4 \frac{\text{kg m}}{\text{s}}$$

The impulse to change the velocity of the automobile is given by the difference in the momentum

$$I = q_f - q_i = 3 \times 10^4 \frac{\text{kg m}}{\text{s}} - 0.75 \times 10^4 \frac{\text{kg m}}{\text{s}}$$
$$= 2.25 \times 10^4 \frac{\text{kg m}}{\text{s}}$$

(E4.4) Given three material points of masses $m_1 = 3.2$ kg, $m_2 = 4.7$ kg, and $m_3 = 5.3$ kg, having, with respect to a Cartesian axis system, coordinates $P_1 = (-1; 0)$, $P_2 = (0; 2)$, and $P_3 = (1; -1)$, respectively. The coordinates of the center of mass are.

(A) (0.25; 1.25)
(B) (−0.64; 2)
(C) (0.16; 0.31) (correct answer)
(D) (0.55; −0.42)
(E) (−0.45; 0.76).

Exercises

(4.1) A body of mass $m_1 = 3$ kg is thrown with velocity $v_1 = 8$ m/s on a rough plane with dynamic friction coefficient $\mu_d = 0.2$ against another mass $m_2 = 2$ kg at distance $d_1 = 60$ cm from the first, against which it bumps completely inelastically. At distance $d_2 = 30$ cm from m_2, there is a spring of elastic constant $k = 100$ N/m. How fast do the masses begin to move after the impact (m/s)?

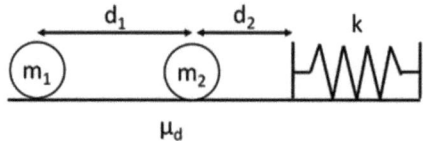

(A) 5.25
(B) 4.71
(C) 6.23
(D) 8.45
(E) 4.80.

(4.2)) By how much does the spring compress in problem 4.1?

(A) 0.45 cm
(B) 1.3 cm
(C) 0.56 m
(D) 2.41 m
(E) 0.93 m.

(4.3) A capital of mass 150 kg laid on a column explodes into 2 pieces: 50 kg falls 3 m north of the column. Where does the remainder of the capital fall?

(A) 6 m north
(B) 1.5 m south
(C) 1.5 m north
(D) 6 m south
(E) 3 m east.

(4.4)) A marble with mass $m_1 = 2$ Kg is launched along a smooth plane where it collides in a completely inelastic manner with a mass $m_2 = 3$ Kg at the foot a rough ramp of length $L = 2$ m inclined at an angle $\alpha = 45°$ to it. Knowing that the dynamic friction coefficient between the ramp and the marble is worth $\mu_d = 0.2$, determine the minimum speed of the marble m_1 to travel the entire ramp.

(A) 30.74 m/s
(B) 11.45 m/s
(C) 24.56 m/s

4.1 Collisions

(D) 16.45 m/s
(E) 14.43 m/s.

(4.5) For an inelastic shock, what is not true?

(A) Momentum is conserved
(B) Some of the energy can be converted into heat
(c) Part of the energy can be converted into acoustic waves
(D) Mechanical energy is conserved
(E) Bodies, by colliding, can deform themselves.

(4.6) The principle of conservation of mechanical energy states that

(A) the change in mechanical energy is due only to the work of non-conservative forces
(B) the change in mechanical energy is due only to the work of conservative forces
(C) the change in mechanical energy is due to the work of both non-conservative and conservative forces
(D) the change in mechanical energy is due only to the elastic force, weight force, and gravitational interaction
(E) the change in kinetic energy is equal to the work done by the forces, provided that the potential energy is zero.

Chapter 5
Gravity

5.1 Kepler's Laws

(1) The planets describe elliptical orbits in which the Sun occupies one of the two foci.
(2) The vector radius that joins the center of the Sun to the center of the planet describes equal areas at equal times (i.e., the areolar velocity, in m²/s, of the planets is constant).
(3) The squares of the periods of revolution are proportional to the cubes of the major semi-axes of the orbits:

$$T^2 = ka^3 \tag{5.1}$$

5.2 Determination of the gravitational interaction

From Kepler's laws, we can derive the trend of the forces of gravitational attraction as a function of the distance between the masses.

Let us make two approximations:

(1) Consider a planet describing a circular orbit of radius R. We can see a circular orbit as a special case of an elliptical orbit in which the semi-axes, major and minor, are equal. Given the small eccentricity of the elliptical orbit of planets in the solar system, the approximation is acceptable.
(2) We limit ourselves to the two-body Sun-Earth system. The solar system includes many more bodies (Sun star, eight planets with satellites, nano-planets like Pluto, Eris, Haumea, etc., and many other smaller bodies), but the interaction, between Earth and, for example, Moon, is much weaker than the interaction between Earth and Sun. Therefore, the second approximation is also acceptable.

By Newton's second law, it turns out that the planet moves with constant velocity as it travels along its orbit (uniform circular motion, in which the contribution of the acceleration is only normal to the Earth's motion):

$$\left|\vec{F}\right| = ma_N = m\frac{v^2}{R} \tag{5.2}$$

Recalling the relation between velocity and period in the uniform circular motion:

$$v = \frac{2\pi R}{T} \rightarrow F = m\frac{v^2}{R} = m\frac{4\pi^2 R}{T^2} \tag{5.3}$$

By Kepler's third law ($T^2 = kR^3$, where k is a constant of proportionality):

$$F = m\frac{4\pi^2 R}{T^2} = m4\pi^2 R \frac{1}{kR^3} = m\frac{4\pi^2}{kR^2} \tag{5.4}$$

Let us apply the third principle of dynamics to a system composed of two bodies (e.g., Sun-Earth):

- Force acting on Earth due to the Sun's mass:

$$F_{T,S} = m_T \frac{4\pi^2}{k_T R^2} \tag{5.5}$$

- Force acting on Sun due to the Earth's mass:

$$F_{S,T} = m_S \frac{4\pi^2}{k_S R^2} \tag{5.6}$$

From the third principle of dynamics, it is known that these two forces in modulus are equal:

$$F_{T,S} = F_{S,T} \rightarrow m_T k_S = m_S k_T \tag{5.7}$$

It follows that the product $m_i k_j$ is constant for each pair of bodies i and j.
By introducing a constant γ such that

$$\gamma = \frac{4\pi^2}{m_T k_S} = \frac{4\pi^2}{m_S k_T} = \frac{4\pi^2}{m_i k_j} \tag{5.8}$$

it is possible to write the law of gravitational interaction (proposed by Newton in 1687):

5.2 Determination of the gravitational interaction

$$F = \gamma \frac{m_i m_j}{r_{ij}^2} \tag{5.9}$$

Knowing that centripetal acceleration is directed toward the Sun, which is at the center of the circumference, we can say that the vector $\vec{F}$ describes an attractive force directed along the conjunction between the two bodies. Therefore

$$\vec{F} = -\gamma \frac{m_i m_j}{r_{ij}^2} \hat{u}_r \tag{5.10}$$

So the gravitational force (or interaction) is directly proportional to the masses of the interacting bodies and inversely proportional to the square of the distance between the interacting bodies:

$\gamma = 6.67 \times 10{-11}$ Nm^2/Kg^2 is the universal gravitation constant (measured by Cavendish in 1797–1798). In many textbooks, the universal gravitation constant is denoted by the symbol "G."

Food for thought: Two anomalies of Newton's law of gravitational interaction have been experienced in the history of science, leading to two opposing outcomes. The first anomaly concerns inaccuracies in describing the orbit of Uranus. Such inaccuracies, instead of falsifying the validity of Newton's theory, led to the discovery of Neptune (1845). The second anomaly concerns Mercury's perihelion (closest point of the planet to the Sun), which is closer than predicted. This anomaly supported Einstein's formulation of general relativity: a shorter perihelion of Mercury can be explained by a greater curvature of space-time close to a large mass like that of the Sun.

To confirm the validity of the gravitational interaction also in the case of elliptical orbits, it is useful to introduce the angular momentum:

$$\vec{L} = \vec{\Omega P} \times \vec{q} = \vec{r} \times \vec{q} = \vec{r} \times m\vec{v} \tag{5.11}$$

where $\vec{\Omega P}$ is the distance between a pole Ω and the point P.

The torque is given by (Euler's second law)

$$\vec{M} = \frac{d}{dt}\vec{L} = \frac{d\vec{r}}{dt} \times m\vec{v} + \vec{r} \times m\frac{d\vec{v}}{dt} \tag{5.12}$$

For fixed pole the first term is zero, therefore

$$\frac{d}{dt}\vec{L} = \vec{r} \times \frac{d(m\vec{v})}{dt} = \vec{r} \times \frac{d\vec{q}}{dt} = \vec{r} \times \vec{F} = \vec{M} \tag{5.13}$$

Food for thought: In the case of a mobile pole, the material point P is defined by the vector $\vec{r}$ from the pole O (fixed) and by the vector $\vec{r}\prime$ from the pole O' (mobile). The position vector of O' with respect to O is $\vec{r}_{O\prime}$

$$\vec{r} = \vec{r}_{O\prime} + \vec{r}\prime \tag{5.14}$$

$$\vec{r}\prime = \vec{r} - \vec{r}_{O\prime} \tag{5.15}$$

$$\vec{L}_{O\prime} = \vec{r}\prime \times m\vec{v} = (\vec{r} - \vec{r}_{O\prime}) \times m\vec{v} \tag{5.16}$$

Derivate of $\vec{L}_{O\prime}$:

$$\begin{aligned}
\frac{d\vec{L}_{O\prime}}{dt} &= \frac{d}{dt}[(\vec{r} - \vec{r}_{O\prime}) \times m\vec{v}] \\
&= \frac{d(\vec{r} - \vec{r}_{O\prime})}{dt} \times m\vec{v} + (\vec{r} - \vec{r}_{O\prime}) \times \frac{d(m\vec{v})}{dt} \\
&= \frac{d\vec{r}}{dt} \times m\vec{v} - \frac{d\vec{r}_{O\prime}}{dt} \times m\vec{v} + (\vec{r} - \vec{r}_{O\prime}) \times m\vec{a} \\
&= \vec{v} \times m\vec{v} - \frac{d\vec{r}_{O\prime}}{dt} \times m\vec{v} + (\vec{r} - \vec{r}_{O\prime}) \times \vec{F} \\
&= -\frac{d\vec{r}_{O\prime}}{dt} \times m\vec{v} + \vec{M}_{O\prime}
\end{aligned} \tag{5.17}$$

Thus

$$\frac{d\vec{L}_{O\prime}}{dt} = -\frac{d\vec{r}_{O\prime}}{dt} \times m\vec{v} + \vec{M}_{O\prime} \tag{5.18}$$

Central force: The pole coincides with the center of force. The areolar velocity is given by $\sigma = dA/dt$.

The modulus of the angular momentum can be written as

$$L = mr^2 \frac{d\vartheta}{dt} = mr^2 \frac{2dA}{r^2 dt} = \frac{2m\, dA}{dt} \tag{5.19}$$

In the case of central force:
$\sigma = $ constant $\to \vec{L} = $ constant
[Alternative demonstration:

$$\vec{M} = \vec{r} \times \vec{F} = |\vec{r}||\vec{F}|\hat{n} \sin\alpha_{rF} \tag{5.20}$$

Since $\vec{F}$ is a central force, the angle between and is 0°, so the sine of the angle is zero. So $\vec{M} = \frac{d\vec{L}}{dt} = 0 \to \vec{L} = $ costant.]

5.3 Potential Energy for Gravitational Interaction

It can be easily seen that the work of the gravitational interaction does not depend on the trajectory. The work is

$$L_{AB} = -\Delta E_p \qquad (5.21)$$

The difference in potential energy is

$$\Delta E_p = \frac{\gamma m_1 m_2}{r_1} - \frac{\gamma m_1 m_2}{r_2} \qquad (5.22)$$

$$If\, r_1 \to \infty,\, \Delta E_p = -\frac{\gamma m_1 m_2}{r}$$

Food for thought: With the conservations of the angular momentum and the mechanic energy

$$\begin{cases} \vec{L} = \vec{r} \times \vec{q} = \vec{r} \times m\vec{v} \\ E = E_k + E_p = \frac{1}{2}mv^2 - \gamma \frac{mM}{r} \end{cases} \qquad (5.23)$$

it is possible to derive a more precise formulation of the potential energy, i.e., the effective potential energy. The mechanic energy can be written as

$$E = E_{k,\text{radial}} + E_{p,\text{effective}} \qquad (5.24)$$

with

$$E_{k,\text{radial}} = \frac{1}{2}m\dot{r}^2 \qquad (5.25)$$

$$E_{p,\text{effective}} = \frac{L^2}{2mr^2} - \gamma \frac{mM}{r} \qquad (5.26)$$

In the figure, the black curve is the effective potential energy, while the red curve is the potential energy.

The effective potential energy has a vertical asymptote near the origin, an area of positive concavity with a point of minimum, an ascent as r increases tending asymptotically to zero (Fig. 5.1).

The orbit of a body subject to central force inversely proportional to the square of distance can be described by a conic function:

$$(1 - \varepsilon^2)x^2 + 2\varepsilon\rho x + y^2 = \rho^2 \qquad (5.27)$$

Fig. 5.1 Potential energy of the gravitational interaction (red curve) and effective potential energy of the gravitational interaction (red curve).

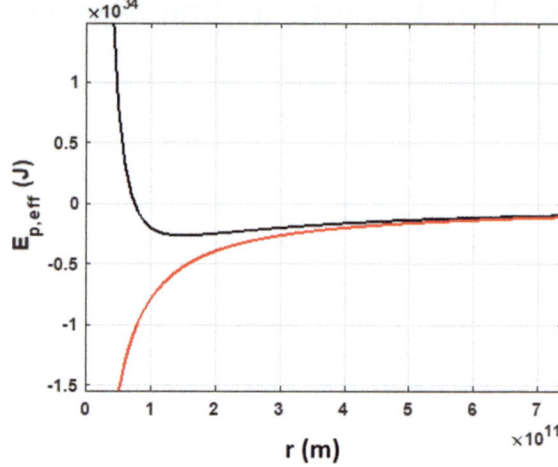

with

$$\rho = \frac{L^2}{mk}; \varepsilon = \sqrt{1 + \frac{2EL^2}{mk^2}};$$

$$k = \gamma mM; E = E_k + E_p \tag{5.28}$$

The type of orbit depends on the eccentricity ε of the conic function:

- $\varepsilon > 1$: Hyperbolic orbit;
- $\varepsilon = 1$: Parabolic orbit;;
- $0 \leq \varepsilon < 1$: Elliptical orbit.

With $\varepsilon = 0$, the conic function is simply $x^2 + y^2 = \rho^2$, i.e., a circumference.

5.4 Fundamental Interactions

Four fundamental interactions have been identified to explain phenomena in nature:

(1) Gravitational interaction: attraction between(masses.
(2) Electromagnetic interaction: electric charges repelling each other (same sign) or attracting each other (opposite sign). Protons are characterized by an irreducible amount of positive charge $+q = 1,6 \times 10^{-19} C$, , C stands for Coulomb, a unit of charge measurement), while electrons are characterized by an irreducible amount of negative charge $(-q = -1,6 \times 10^{-19} C)$.

(3) Weak interaction: interaction between sub-atomic particles responsible for the radioactive decay of atoms.
(4) Strong interaction: interaction that holds quarks together in protons and neutrons and binds protons and neutrons in nuclei.

In recent years, the last three interactions, electromagnetic, weak, and strong have been unified with the standard model. However, it is not yet possible to find a model that takes into account all four interactions simultaneously.

One reason for this is the "weakness" of the gravitational interaction compared to the others.

5.5 Comparison of Gravitational Force and Electrostatic Force

A simple comparison is between the attractive electrostatic force between a proton and a neutron, and the gravitational force between the masses of the proton and neutron, in the hydrogen atom. The hydrogen atom can be schematized very simply with a microscopic solar system in which the proton in place of the Sun and the electron as a planet revolve around the Sun traveling in a circular trajectory.

[In reality, the electron does not travel in a circular trajectory, but moves with a random motion in a region of space around the nucleus. That region is called an orbital and can be determined by solving the Schrödinger equation for the hydrogen atom.]

The attraction between the proton and the electron is determined by Coulomb's law:

$$\vec{F}_e = \frac{1}{4\pi\varepsilon_0} \frac{q(-q)}{r^2} \hat{u}_r \qquad (5.29)$$

Note that the formula is very similar to that of gravitational interaction. In this case, the force is attractive because the proton and electron charges are of opposite sign.

Data for the hydrogen atom (proton + electron):

m_e (electron mass): 9×10^{-31} Kg

m_p (proton mass): 1.7×10^{-27} Kg.

Nucleons and electrons have both charge $e = 1,6 \times 10^{-19} C$, but with the opposite sign.

The vacuum dielectric constant (or dielectric permittivity) is $\varepsilon_0 = 8.85 \times 10^{-12} C^2/Nm^2$.

The moduli of the two forces are

$$F_e = \frac{q^2}{4\pi\varepsilon_0 r^2}; \quad F_g = \gamma \frac{m_p m_e}{r^2} \qquad (5.30)$$

And their ratio is

$$\frac{F_{elettrostatica}}{F_{gravitazionale}} = \frac{q^2}{4\pi\varepsilon_0 \gamma m_p m_e} \cong 2.3 \times 10^{39} \qquad (5.31)$$

It is inferred that matter is held together solely by electrical forces.

In general, an estimate of the intensity ratios among the various interactions tells us that at the level of distances between bodies of 10^{-15} m (1 femtometer, fm) the strong interaction is about 100 times more intense than the electromagnetic interaction, a million times more intense than the weak interaction, and about 10^{41} times more intense than the gravitational interaction.

5.6 Examples

(E5.1) The square of the semi-major axis of the orbit of planet 1 is twice the square of the semi-major axis of the orbit of planet 2. The ratio (period T_1 of revolution of planet 1)/(period T_2 of revolution planet 2) is

$$a_1^2 = 2a_2^2$$

$$a_1 = \sqrt{2}a_2$$

$$T_1^2 = ka_1^3 = k2^{3/2}a_2^3$$

$$T_2^2 = ka_2^3$$

$$\frac{T_1^2}{T_2^2} = \frac{k2^{3/2}a_2^3}{ka_2^3} = 2^{3/2}$$

$$\frac{T_1}{T_2} = 2^{\frac{3}{2}\cdot\frac{1}{2}} = 2^{\frac{3}{4}} = 1.682$$

(E5.2) For a 50 kg satellite at an altitude of 100 km, what are the velocity and period of revolution?

Solution

To find the velocity, it is possible to use the second principle of dynamics:

$$\gamma \frac{m_{Earth} m_{satellite}}{r^2} = ma_N = m_{satellite} \frac{v^2}{r}$$

5.6 Examples

Thus

$$v = \sqrt{\gamma \frac{m_{\text{Earth}}}{r}}$$

The radius r is the sum of the Earth's radius and the satellite altitude, thus, 6371 km + 100 km = 6471 km.

In this case

$$v = \sqrt{6.67 \times 10^{-11} \, \text{Nm}^2/\text{kg}^2 \frac{5.97 \times 10^{24} \, \text{kg}}{6471000 \, \text{m}}} = 7845 \, \text{m/s}$$

The period of revolution is

$$T = \frac{2\pi r}{v} = \frac{2\pi \, 6471000 \, \text{m}}{7845 \, \text{m/s}} = 5183 \, \text{s}$$

Thus, the satellite makes a loop around the Earth in 1 h, 26 min, and 23 s.

(E5.3) Sunlight takes 1 h, 19 min, and 27 s to reach Saturn. Knowing that the Sun has a mass of 1.98×10^{30} kg, find the planet's period of revolution.

Solution

1 h, 19 min, and 27 s corresponds to 4767 s.

In this time, considering the speed of light $c = 3 \times 10^8$ m/s

$$s = a = vt = ct$$
$$= (3 \times 10^8 \, \text{m/s})(4767 \, \text{s}) = 1.43 \times 10^{12} \, \text{m}$$

where a is the major semi-axis of Saturn's orbit.

Kepler's third law states that $T^2 = ka^3$.

To find k, it is possible to use

$$\gamma = \frac{4\pi^2}{m_{\text{Sun}} k}$$

Thus

$$k = \frac{4\pi^2}{m_{\text{Sun}} \gamma}$$

$$= \frac{4\pi^2}{(1.98 \times 10^{30} \, textkg)(6.67 \times 10^{-11} \, \text{Nm}^2/\text{kg}^2)} \& = 2.98 \times 10^{-19} \, \text{s}^2/\text{m}^3$$

Using Kepler's third law, it is possible to find T

$$T^2 = ka^3$$

Thus

$$T = \sqrt{ka^3}$$
$$= \sqrt{(2.98 \times 10^{-19} \, texts^2/m^3)(1.43 \times 10^{12} \, m)^3}$$
$$= 9.35 \times 10^8 \, texts$$

Exercises

(5.1) If a planet has a mass that is half that of Earth, to have the same gravitational acceleration it must have

(A) the same radius
(B) a radius twice as large as that of the Earth
(C) a radius 1.414 times smaller than that of the Earth
(D) a radius half the radius of that of the Earth
(E) a gravitational constant half that of Earth's.

(5.2) Saturn has an equatorial diameter of 120536 km and a mass of $5{,}6846 \times 10^{26}$ kg. If we were on the surface of Saturn's equator, how much would the gravitational acceleration be worth?

(A) 375.9% relative to that of the Earth's
(B) 12.8 m/s^2
(C) 26.6% of that of the Earth's
(D) 10.8% of that of the Earth's
(E) 285% of that of the Earth's.

(5.3) Pluto (m = 1.3 x 10^{22} kg) and Charon (m = 1.5 x 10^{21} kg) are 19571 km apart. Earth (m = 6.0 x 10^{24} kg) and Moon (m = 7.3 x 10^{22} kg) are 384400 km apart. Saturn (m=5.7 x 10^{26} kg) and Hyperion (m = 5.6 x 10^{18} kg) are 1500934 km apart. Order the planet-satellite gravitational interactions by intensity:

(A) P/C, T/L, S/I
(B) T/L, P/C, S/I
(C) S/I, T/L, P/C
(D) S/I, P/C, T/L
(E) P/C, S/I, T/L.

(5.4) The gravitational acceleration is determined as

(A) $g = \gamma \frac{M_{Earth}}{r_{Earth}^2}$
(B) $g = \gamma \frac{r_{Earth}^2}{M_{Earth}}$
(C) $g = \gamma \frac{M_{Earth}}{r_{ext Earth}}$
(D) $g = \gamma \frac{r_{Earth}}{M_{Earth}}$
(E) $g = \gamma^2 \frac{M_{Earth}}{r_{Earth}}$.

5.6 Examples

(5.5) If the semi-major axis of planet A's orbit is 5 times larger than the semi-major axis of planet B's orbit, what is the ratio of the two periods of revolution (T_A/T_B) worth?

(A) $5^{5/2}$
(B) $5^{-3/2}$
(C) 125
(D) $5^{-2/5}$
(E) $5\sqrt{5}$.

Chapter 6
Relative Motions

In Fig. 6.1, the reference frame O' is moving with velocity $v_{O'}$ with respect to the reference frame O.

The positions can be written as

$$\vec{r}' = \vec{r} - \vec{r}_{O'} \tag{6.1}$$

$$\vec{r} = \vec{r}' + \vec{r}_{O'} \tag{6.2}$$

In Fig. 6.2, the reference frame O' is moving with an angular velocity $\vec{\omega}$ with the respect to the reference frame O:

The positions can be written as

$$\vec{r} = \vec{r}' \tag{6.3}$$

with

$$\vec{r} = x\hat{u}_x + y\hat{u}_y + z\hat{u}_z \tag{6.4}$$

and

$$\vec{r}' = x'\hat{u}_{x'} + y'\hat{u}_{y'} + z'\hat{u}_{z'} \tag{6.5}$$

Reference frame in **roto-translational** motion
- Relationship between speeds:

$$\vec{v} = \vec{v}' + \vec{v}_{O'} + \left(\vec{\omega} \times \vec{r}'\right) \tag{6.6}$$

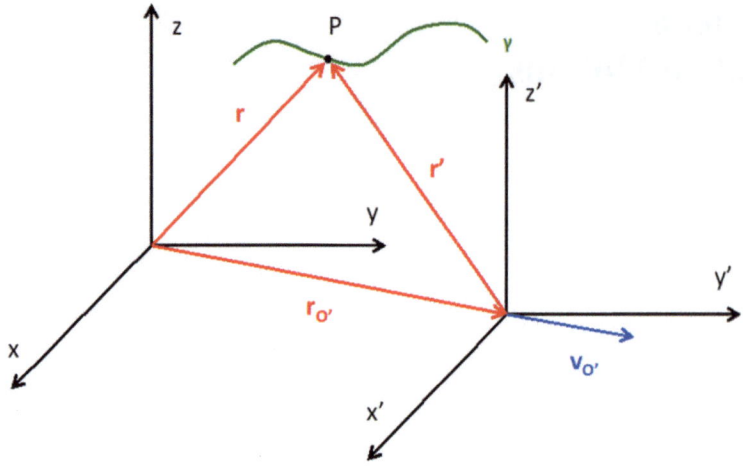

Fig. 6.1 Sketch of the reference frame O', moving with velocity $v_{O'}$ with respect to the reference frame O

Fig. 6.2 Reference frame O' moving with an angular velocity $\vec{\omega}$ with the respect to the reference frame O

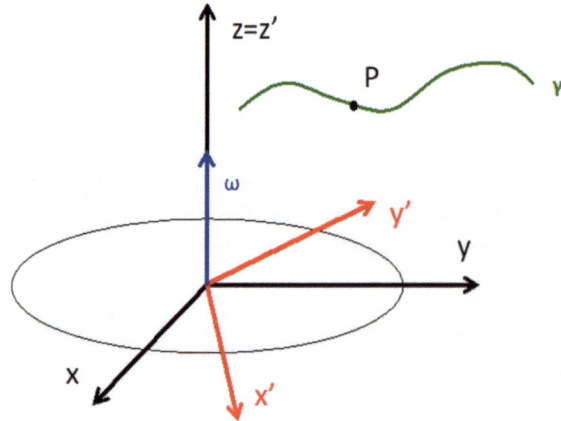

- Relationship between accelerations:

$$\vec{a} = \vec{a}' + \vec{a}_{O'} + 2\vec{\omega} \times \vec{v}' + \vec{\omega} \times (\vec{\omega} \times \vec{r}') \tag{6.7}$$

In an **inertial reference frame** ($\vec{v}_{O'} =$ costante; $\vec{a}_{O'} = 0$; $\omega = 0$) the principle of inertia applies. For all inertial frames, the second law of dynamics $\vec{F} = \frac{d\vec{q}}{dt} = m\vec{a}$ is written in the same way.

Non-inertial reference frame: ($\vec{a}_{O'} \neq 0 e/o \omega \neq 0$), the second law of dynamics is no longer valid. With $\vec{a} = \vec{a}' + \vec{a}_{O'} + 2\vec{\omega} \times \vec{v}' + \vec{\omega} \times (\vec{\omega} \times \vec{r}')$, if $\vec{F} = m\vec{a}$ in the inertial frame, in the mobile system

6 Relative Motions

$$\vec{F} - m\vec{a}_c - m\vec{a}_t = m\vec{a}'$$ (6.8)

where the complementary or Coriolis acceleration: $\vec{a}_c = 2\vec{\omega} \times \vec{v}'$
and drag acceleration: $\vec{a}_t = \vec{a}_{O'} + \vec{\omega} \times (\vec{\omega} \times \vec{r}')$.

In Eq. 6.8, there are **true forces** $\vec{F}$ plus **apparent** forces (or **inertia forces**) $m\vec{a}_c$ and $m\vec{a}_t$.

Examples

(E6.1) A scale is in the elevator and a mass of 20 kg is on the scale. Describe the forces on the non-inertial reference frame attached to the elevator when the elevator is still, when it is accelerating upward with $a = 2m/s^2$, when the velocity of the elevator is constant and $v = 1m/s$, when it is decelerating with $a = -2m/s^2$.

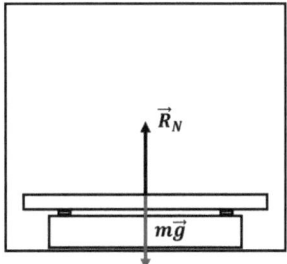

Solution
Weight of a 20 kg object with a stationary elevator

$$\left|\vec{R}_N\right| - \left|\vec{W}\right| = \left|\vec{R}_N\right| - |m\vec{g}| = 0$$

$$\left|\vec{R}_N\right| = \left|\vec{W}\right| = 196N$$

Weight of a 20 kg object with elevator accelerating upward with $a = 2$ m/s²

$$\left|\vec{R}_N\right| - \left|\vec{W}\right| = |m(+\vec{a})|$$

$$\left|\vec{R}_N\right| = \left|\vec{W}\right| + |m(+\vec{a})| = |m\vec{g}| + |m\vec{a}| = 236N$$

Weight of a 20 kg object with elevator traveling at a speed of 1 m/s

$$\left|\vec{R}_N\right| - \left|\vec{W}\right| = m(0)$$

$$|\vec{R}_N| = |\vec{W}| = 196N$$

Weight of a 20 kg object with elevator decelerating with $a = -2$ m/s²

$$|\vec{R}_N| - |\vec{W}| = |m(-\vec{a})|$$

$$|\vec{R}_N| = |\vec{W}| + |m(-\vec{a})| = |m\vec{g}| - |m\vec{a}| = 156N$$

(E6.2) O' is a reference system bound to a train that is accelerating with acceleration a_t in the direction $+x$. In the train a glass is falling down. What is an observer on a reference system O out of the train? And an observer on the accelerating system O'?

Solution

From O:

The glass is falling in the direction $-y$ and the time needed for the glass to touch the ground is

$$t_c = \sqrt{2h/g}$$

The displacement of the glass in the x direction is zero. Within the time t_c the displacement of the train in the x direction is

$$x_{O'} = (1/2)a_t t_c^2$$

When the glass touches the ground, the distance between the glass and the train is

$$distance = (1/2)a_t t_c^2 = (1/2)a_t(2h/g) = a_t h/g$$

From O':

The acceleration of the glass has two components: a_t in the direction $-x'$ and g in the direction $-y'$

$$\vec{a'} = -a_t \hat{u}_x - g\hat{u}_y$$

The observer on O' sees an oblique trajectory for the glass.

(E6.3) Derivate $\vec{v} = \vec{v'} + (\vec{\omega} \times \vec{r'})$ with the reference frame O' in rotation with respect to the reference frame O.

6 Relative Motions

Solution

$$\vec{v} = \frac{d\vec{r}}{dt} = \frac{d\vec{r}'}{dt} = \frac{d}{dt}(x'\hat{u}_{x'} + y'\hat{u}_{y'} + z'\hat{u}_{z'})$$

$$= \frac{dx'}{dt}\hat{u}_{x'} + \frac{dy'}{dt}\hat{u}_{y'} + \frac{dz'}{dt}\hat{u}_{z'} + x'\frac{d\hat{u}_{x'}}{dt} + y'\frac{d\hat{u}_{y'}}{dt} + z'\frac{d\hat{u}_{z'}}{dt}$$

The first three terms are $\vec{v}' = \frac{dx'}{dt}\hat{u}_{x'} + \frac{dy'}{dt}\hat{u}_{y'} + \frac{dz'}{dt}\hat{u}_{z'}$.

For the other three terms Poisson's rule can be used: $\frac{d\hat{u}_{x'}}{dt} = \vec{\omega} \times \hat{u}_{x'}$ (the rule is analogous for $\hat{u}_{y'}$ and $\hat{u}_{z'}$)

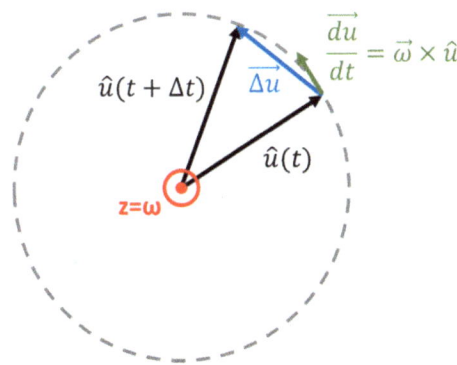

$$\vec{v} = \vec{v}' + x'\left(\vec{\omega} \times \hat{u}_{x'}\right) + y'\left(\vec{\omega} \times \hat{u}_{y'}\right) + z'\left(\vec{\omega} \times \hat{u}_{z'}\right)$$
$$= \vec{v}' + \left(\vec{\omega} \times x'\hat{u}_{x'}\right) + \left(\vec{\omega} \times y'\hat{u}_{y'}\right) + \left(\vec{\omega} \times z'\hat{u}_{z'}\right)$$
$$= \vec{v}' + \vec{\omega} \times (x'\hat{u}_{x'} + y'\hat{u}_{y'} + z'\hat{u}_{z'})$$
$$= \vec{v}' + \vec{\omega} \times \vec{r}'$$

Thus $\vec{v} = \vec{v}' + \vec{\omega} \times \vec{r}'$

(E6.4) Derive $\vec{a} = \vec{a}' + 2\vec{\omega} \times \vec{v}' + \vec{\omega} \times \left(\vec{\omega} \times \vec{r}'\right)$ with the reference frame O' in rotation with respect to the reference frame O.

Solution
The velocity, with the reference frame O' in rotation with respect to the reference frame O, is

$$\vec{v} = \vec{v}' + \left(\vec{\omega} \times \vec{r}'\right)$$

Thus, the acceleration is

$$\vec{a} = \frac{d\vec{v}}{dt} = \frac{d}{dt}\left[\vec{v}\prime + \left(\vec{\omega} \times \vec{r}\prime\right)\right]$$

$$= \frac{d\vec{v}\prime}{dt} + \frac{d}{dt}\left(\vec{\omega} \times \vec{r}\prime\right)$$

$$= \frac{d\vec{v}\prime}{dt} + \left(\frac{d\vec{\omega}}{dt} \times \vec{r}\prime\right) + \left(\vec{\omega} \times \frac{d\vec{r}\prime}{dt}\right)$$

The angular velocity is constant and, consequently, $\frac{d\vec{\omega}}{dt} \times \vec{r}\prime = 0$. Since $\vec{r} = \vec{r}\prime$, then

$$\frac{d\vec{r}\prime}{dt} = \frac{d\vec{r}}{dt} = \vec{v} = \vec{v}\prime + \left(\vec{\omega} \times \vec{r}\prime\right)$$

Thus

$$\vec{a} = \frac{d\vec{v}\prime}{dt} + \left\{\vec{\omega} \times \left[\vec{v}\prime + \left(\vec{\omega} \times \vec{r}\prime\right)\right]\right\}$$

$$= \frac{d\vec{v}\prime}{dt} + \vec{\omega} \times \vec{v}\prime + \vec{\omega} \times \left(\vec{\omega} \times \vec{r}\prime\right)$$

The first term $\frac{d\vec{v}\prime}{dt}$, knowing that $\vec{v}\prime = \frac{dx\prime}{dt}\hat{u}_{x\prime} + \frac{dy\prime}{dt}\hat{u}_{y\prime} + \frac{dz\prime}{dt}\hat{u}_{z\prime}$, is

$$\frac{d\vec{v}\prime}{dt} = \frac{d}{dt}\left(\frac{dx\prime}{dt}\hat{u}_{x\prime} + \frac{dy\prime}{dt}\hat{u}_{y\prime} + \frac{dz\prime}{dt}\hat{u}_{z\prime}\right)$$

$\frac{d}{dt}\left(\frac{dx\prime}{dt}\hat{u}_{x\prime}\right)$ is the derivative of a product. Also $\frac{d}{dt}\left(\frac{dy\prime}{dt}\hat{u}_{y\prime}\right)$ and $\frac{d}{dt}\left(\frac{dz\prime}{dt}\hat{u}_{z\prime}\right)$. Thus

$$\frac{d\vec{v}\prime}{dt} = \frac{d^2x\prime}{dt^2}\hat{u}_{x\prime} + \frac{d^2y\prime}{dt^2}\hat{u}_{y\prime} + \frac{d^2z\prime}{dt^2}\hat{u}_{z\prime} + \frac{dx\prime}{dt}\frac{d\hat{u}_{x\prime}}{dt} + \frac{dy\prime}{dt}\frac{d\hat{u}_{y\prime}}{dt} + \frac{dz\prime}{dt}\frac{d\hat{u}_{z\prime}}{dt}$$

The first three terms are the components of $\vec{a}\prime$. For the other three terms, it is possible to employ Poisson's rule:

$$\frac{d\vec{v}\prime}{dt} = \vec{a}\prime + \frac{dx\prime}{dt}\left(\vec{\omega} \times \hat{u}_{x\prime}\right) + \frac{dy\prime}{dt}\left(\vec{\omega} \times \hat{u}_{y\prime}\right) + \frac{dz\prime}{dt}\left(\vec{\omega} \times \hat{u}_{z\prime}\right)$$

$$= \vec{a}\prime + \left[\vec{\omega} \times \left(\frac{dx\prime}{dt}\hat{u}_{x\prime} + \frac{dy\prime}{dt}\hat{u}_{y\prime} + \frac{dz\prime}{dt}\hat{u}_{z\prime}\right)\right]$$

$$= \vec{a}\prime + \vec{\omega} \times \vec{v}\prime$$

To define the acceleration $\vec{a}$, it is possible to state that

6 Relative Motions

$$\vec{a} = \frac{d\vec{v}'}{dt} + \vec{\omega} \times \vec{v}' + \vec{\omega} \times \left(\vec{\omega} \times \vec{r}'\right)$$

in which

$$\frac{d\vec{v}'}{dt} = \vec{a}' + \vec{\omega} \times \vec{v}'$$

Thus,

$$\vec{a} = \vec{a}' + \vec{\omega} \times \vec{v}' + \vec{\omega} \times \vec{v}' + \vec{\omega} \times \left(\vec{\omega} \times \vec{r}'\right) =$$
$$= \vec{a}' + 2\vec{\omega} \times \vec{v}' + \vec{\omega} \times \left(\vec{\omega} \times \vec{r}'\right)$$

(E6.5) A material point of mass m is resting on the edge of a rotating platform. The system rotates with constant $\vec{\omega}$. The point is held by an inextensible and massless wire at a distance r from the axis of rotation. Describe the motion with the wire and after the wire breaks.

Solution

With the wire:
   On O (circumference center, first figure):

$$\vec{T} = m\vec{a} \Rightarrow \vec{T} = -m\omega^2 r \hat{u}_r$$

On O' (on the material point, non-inertial system, second figure):

$$\vec{T} + \vec{F}_{\text{centrifugal}} = 0 \Rightarrow \vec{F}_{\text{centrifugal}} = m\omega^2 r \hat{u}_r$$

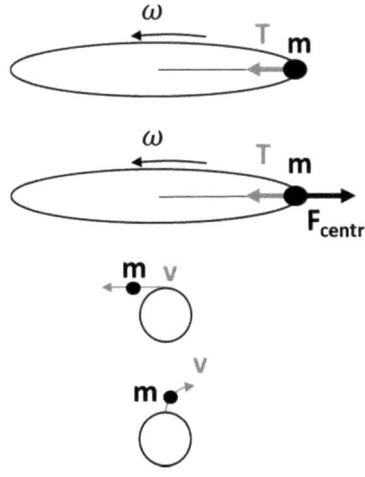

After the wire breaks:
On O (third figure):

$$v = \omega r$$

(only tangent velocity)
On O' (fourth figure):

$$\vec{F}_{centrifugal} + \vec{F}_{Coriolis} = m\vec{a}' = m\omega^2 r \hat{u}_r - 2m\vec{\omega} \times \vec{v}'$$

Centrifugal force and Coriolis force occur.

Exercises

(6.1) In a reference system O a marble falls from a height h along -z. What does a second system O', moving with constant velocity v in a direction parallel to the x-axis of O, see?

(A) the marble falls at a distance $-h$ from O'
(B) the marble falls at a distance $-v\sqrt{2g/h}$ from O'
(C) the marble falls at a distance h from O'
(D) the marble falls in the same position of O'
(E) the marble falls at a distance h from O

(6.2) Achilles with shield runs at 10 m/s. To avoid getting hit by an arrow, Achilles puts his shield at 45° to the vertical (direction along which the arrow is coming). The speed of the arrow relative to Achilles is

(A) 10 m/s
(B) 20 m/s
(C) 15.78 m/s
(D) 9.1 m/s
(E) 14.14 m/s

(6.3) An elevator accelerates and decelerates with equal modulus. A man weighs himself in these two elevator regimes and weighs 900 N and 700 N, respectively. What is the mass in kg of the man worth?

(A) 63.3 kg
(B) 78.5 kg
(C) 81.6 kg
(D) 99 kg
(E) 112.6 kg

Chapter 7
Rigid Body

The rigid body is defined as a system of material points in which the mutual distance between any two points remains constant during motion.

The three main properties of the rigid body are

(i) center of mass (CM), fixed with respect to the other points of the rigid body

$$\vec{r}_{cm} = \frac{\sum_{i=1}^{N} m_i \vec{r}_i}{\sum_{i=1}^{N} m_i} \tag{7.1}$$

(ii) 6° of freedom: Three degrees of freedom are related to translation, three degrees of freedom are related to rotation;
(iii) internal forces do not do work.

The integral calculation of CM is

$$\vec{r}_{cm} = \frac{1}{M} \int \vec{r}\, dm \tag{7.2}$$

if the rigid body is homogeneous (volumetric mass density $\rho = \frac{dm}{dV} = constant$):

$$\vec{r}_{cm} = \frac{1}{V} \int_V \vec{r}\, dV \tag{7.3}$$

In this case, the position of the center of mass depends only on the shape of the object.

For three-dimensional rigid bodies: Volumetric mass density $\rho = dm/dV$
For two-dimensional rigid bodies: Superficial mass density $\sigma = dm/dS$
For one-dimensional rigid bodies: Linear mass density $\lambda = dm/dl$

For the rigid body (divided in N points) the kinetic energy theorem states that

$$E_k = \sum_{i=1}^{N} E_{k,i} = \sum_{i=1}^{N} \frac{1}{2} m_i v_i^2$$

$$= \frac{1}{2}\left(\sum_{i=1}^{N} m_i r_i^2\right)\omega^2 = \frac{1}{2} I \omega^2 \qquad (7.4)$$

where I is the momentum of inertia (MI) with respect to the axis z:

$$I_z = \sum_{i=1}^{N} m_i r_i^2 \qquad (7.5)$$

The integral calculation of MI is

$$I_z = \int r^2 dm \qquad (7.6)$$

Some examples of calculation of the momentum of inertia are given in E7.1, E7.2, E7.3.

MI in several homogeneous solids of mass M (z is axis through CM and axis of symmetry):

Ring: $I_z = MR^2$	Thin disk: $I_z = \frac{1}{2}MR^2$	Sphere: $I_z = \frac{2}{5}MR^2$
Hollow sphere: $I_z = \frac{2}{3}MR^2$	Hollow cylinder: $I_z = MR^2$	Cylinder: $I_z = \frac{1}{2}MR^2$

Food for thought: It is clear that the examples mentioned above are symmetric bodies. For non-symmetric bodies, a numerical calculation of the integral $I_z = \int r^2 dm$ can be necessary.

The Huygens-Steiner (parallel axes) theorem states

$$I = I_{cm} + Md^2 \qquad (7.7)$$

where I is the MI for a rotation around a new axis that is parallel to the one passing through CM).

Proof

Consider a rigid body consisting of N material points and calculate its moment of inertia with respect to any z' axis. We take an xyz Cartesian reference system with the origin in the center of mass cm of the body and the z-axis parallel to z'. Point O is located at the intersection of z' and the xy plane and is at distance d from the center of mass (and the z-axis) (Fig. 7.1).

7 Rigid Body

Fig. 7.1 Cartesian reference system with the origin in the center of mass cm of the body and the z-axis parallel to z'. Point O is located at the intersection of z' and the xy plane and is at distance d from the center of mass

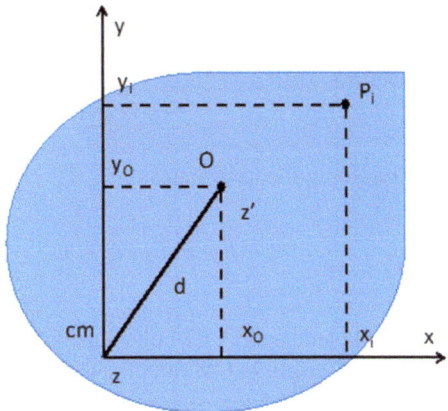

$$d = \sqrt{x_O^2 + y_O^2} \tag{7.8}$$

The momentum of inertia of the rigid body with respect to z' is

$$I_O = \sum_{i=1}^{N} m_i \left[(x_i - x_O)^2 + (y_i - y_O)^2 \right] \tag{7.9}$$

Thus

$$I_O = \sum_{i=1}^{N} m_i \left(x_i^2 + y_i^2 \right) + \left(x_O^2 + y_O^2 \right) \sum_{i=1}^{N} m_i \\ - 2x_O \sum_{i=1}^{N} m_i x_i - 2y_O \sum_{i=1}^{N} m_i y_i \tag{7.10}$$

For the products $m_i x_i$ and $m_i y_i$ it is possible to state

$$\begin{cases} \sum_{i=1}^{N} m_i x_i = M x_{CM} \\ \sum_{i=1}^{N} m_i y_i = M y_{CM} \end{cases} \text{with} \begin{cases} x_{CM} = 0 \\ y_{CM} = 0 \end{cases} \tag{7.11}$$

Thus

$$2x_O \sum_{i=1}^{N} m_i x_i = 0; \; 2y_O \sum_{i=1}^{N} m_i y_i = 0 \tag{7.12}$$

With $d^2 = x_O^2 + y_O^2$ and $\sum_{i=1}^{N} m_i = M$.
Finally, I_O is

$$I_O = I_{z'} = \sum_{i=1}^{N} m_i \left(x_i^2 + y_i^2 \right) + Md^2$$
$$= I_{CM} + Md^2 \qquad (7.13)$$

The König theorem states

$$E_K = E_{k,CM} + E_{k'} = \frac{1}{2} M v_{CM}^2 + \frac{1}{2} I_{CM} \omega^2 \qquad (7.14)$$

[sum of kinetic energy CM plus kinetic energy that the rigid body has in a reference frame with origin in CM (and fixed orientation with respect to the inertial reference frame)].

Rigid body dynamics

The *translation* of a rigid body follows Euler's first law:

$$\vec{F}^{external} = \frac{d\vec{Q}}{dt} = M \vec{a}_{CM} \qquad (7.15)$$

Instead, when the rigid body is *rotating with respect to a fixed axis*, the dynamics should be carefully evaluated.

For the point P_i of the rigid body as in Fig. 7.2:

$$\vec{L}_i = \vec{r}_i \times m_i \vec{v}_i \qquad (7.16)$$

With the scalar relation $v_i = d_i \omega$:

Fig. 7.2 Angular moment $\vec{L}_i$ for the point P_i which is part of the rigid body. The position of the point P_i is $\vec{r}_i$ from the origin O

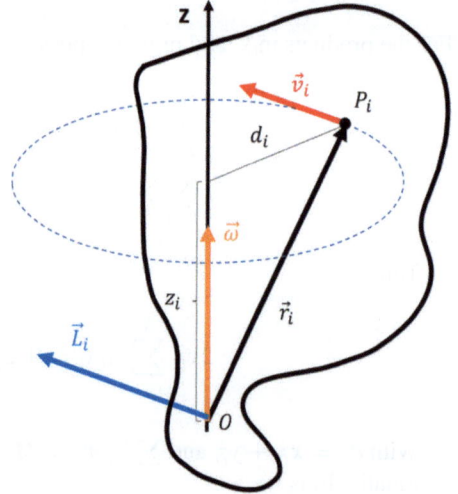

$$\vec{L}_i = \vec{r}_i \times m_i \vec{v}_i = r_i m_i d_i \omega \hat{u}_L \tag{7.17}$$

Thus, the axial angular momentum is

$$\vec{L}_i^z = L_i sen\vartheta_i \hat{u}_z = r_i m_i d_i \omega sen\vartheta_i \hat{u}_z \tag{7.18}$$

And the radial angular momentum:

$$L_i^{rad} = L_i cos\vartheta_i \hat{u}_r = r_i m_i d_i \omega cos\vartheta_i \hat{u}_r \tag{7.19}$$

With $\vec{L}_i = \vec{L}_i^z + \vec{L}_i^{rad}$. With $r_i sen\vartheta_i = d_i$:

$$L_i^z = m_i d_i^2 \omega \tag{7.20}$$

With $r_i cos\vartheta_i = z_i$:

$$L_i^{rad} = m_i d_i \omega z_i \tag{7.21}$$

Finally, the total angular momenta for the rigid body are

(i) axial angular momentum: $L^z = \left(\sum_{i=1}^{N} m_i d_i^2\right)\omega = I_z \omega$;
(ii) radial angular momentum: $L^{rad} = \omega \sum_{i=1}^{N} m_i d_i z_i$.

Zero radial component: (i) rigid plane body that rotates always remaining in the same plane (with pole in the plane, therefore zero z component); (ii) axis of rotation coinciding with an axis of symmetry (the radial components cancel each other).

Thus, there can be two components of the angular momentum, on the axial direction (i.e., parallel to the direction of the rotation axis) and on the radial direction (i.e., on the plane that is normal to the rotation axis).

Consequently, *Euler's second law* can be written for an axial torque and a radial torque:

(i) axial component: $\vec{M}_z = \frac{d\vec{L}_z}{dt} = I_z \frac{d^2\phi}{dt^2} \hat{u}_z = I_z \frac{d\vec{\omega}}{dt} = I_z \vec{\alpha}$;
(ii) radial component: $\vec{M}_{rad} = \frac{d\vec{L}_{rad}}{dt}$.

The torque for the weight force applied to a rigid body is

$$\vec{M} = \vec{r}_{cm} \times m\vec{g} \tag{7.22}$$

which is the moment of the weight force on the center of mass.

For a physical pendulum, i.e., a rigid body attached to a pole with the center of mass at distance d from the pole, it is possible to write

$$M = I_O \alpha = I_O \frac{d^2\vartheta}{dt^2} = -mgd sin\vartheta \tag{7.23}$$

For small oscillations ($\sin\vartheta \cong \vartheta$), it is possible to write the harmonic motion equation: $\frac{d^2\vartheta}{dt^2} + \omega^2\vartheta = 0$ (period: $T = 2\pi\sqrt{\frac{I_O}{mgd}}$).

Static equilibrium: a rigid body is in static equilibrium if the results of external forces and moments of external forces are null and at least in an instant all points of the rigid body are at rest.

Important: the calculation of the moment of the forces in the static case does not depend on the choice of the pole.

Energy in case of rotation variation for the rigid body:

$$dL = \sum_{i=1}^{N} \vec{F}_i^{ext.} \cdot d\vec{s}_i = M_z^{ext.} d\vartheta$$
$$= I_z \omega d\omega = I_z \alpha d\vartheta \qquad (7.24)$$

Thus, the integral work is $L = \int_{\vartheta_1}^{\vartheta_2} M_z^{ext.} d\vartheta$

And the developed instantaneous power is

$$P = \frac{dL}{dt} = M_z^{ext.} \frac{d\vartheta}{dt} = M_z^{est.} \omega \qquad (7.25)$$

Rolling:

(i) the center of mass (CM) translates on a straight line;
(ii) the speed of point A (of the rigid body) combination of the speed of CM and relative velocity in the reference system of CM: $\vec{v}_A = \vec{v}_{CM} + \vec{\omega} \times \vec{r}_A$ ($\vec{\omega}$ equal for all points of rigid body).

Pure rolling: P point of contact with the plane stopped moment by instant in the inertial frame (there is no creeping): $\vec{v}_P = 0 \rightarrow \vec{v}_{CM} = -\vec{\omega} \times \vec{r}_P$ ($\vec{r}_P$ vector radius between CM and contact point P).

Conditions for pure rolling (static friction required, Fig. 7.3):

(i) forces along x: $F + F_a = ma_{cm}$ and y: $R_N - mg = 0$;
(ii) moments: $F_a R - M = I_{cm}\alpha = -I_{cm} a_{cm}/R$.

We find

$$F_a = \frac{mMR - FI_{cm}}{I_{cm} + mR^2} \; ; \; a_{cm} = \frac{FR^2 + MR}{I_{cm} + mR^2} \qquad (7.26)$$

Only force F:

$$F \leq \mu_s mg \left(1 + \frac{mR^2}{I_{cm}}\right) \qquad (7.27)$$

Fig. 7.3 Sketch of pure rolling with force F on the center of mass and moment M

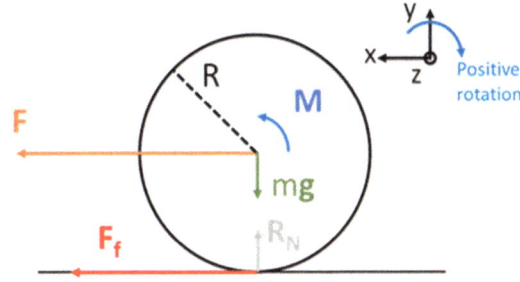

(If greater there is creeping).
M moment only:

$$M \leq \frac{\mu_s g}{R}(I_{cm} + mR^2) \qquad (7.28)$$

(If more slippage occurs).
In pure rolling the system is conservative. In the case of a cylinder:

$$E_k = \frac{1}{2}mv_{cm}^2 + \frac{1}{2}I_{cm}\omega^2$$
$$= \frac{1}{2}(I_{cm} + md^2)\omega^2 = \frac{3}{4}md^2\omega^2 \qquad (7.29)$$

Real case: there is rolling friction:

$$\vec{M}_{av} = \delta \vec{R}_N; \quad \vec{F}_{av} = \delta \vec{R}_N/R \qquad (7.30)$$

(δ distance $\vec{R}_N - CM$ along x).
The rolling friction can be also written as

$$\vec{F}_{av} = \mu_v \vec{R}_N \qquad (7.31)$$

With $\mu_v = \delta/R$, which is the rolling friction coefficient and is unitless.

Examples

(E7.1) Determine the momentum of inertia of a homogeneous ring, with radius R and mass M, with respect to the z-axis that passes through the ring center and is orthogonal to the ring plane.

Solution

The ring can be divided into infinitesimal components with mass dm. The distance between the z-axis and the components is R.

$$I_z = R^2 \int dm = MR^2$$

Alternately, the linear mass density can be used: $\lambda = \frac{M}{2\pi R}$
Each mass component has mass $dm = \lambda dl$. Thus

$$I_z = \int R^2 dm = \int R^2 \lambda dl = R^2 \lambda \int dl = 2\pi \lambda R^3$$

Considering that λ is the linear mass element

$$I_z = 2\pi \lambda R^2 = 2\pi \frac{M}{2\pi R} R^3 = MR^2$$

(E7.2) Determine the momentum of inertia of a homogeneous disk, with radius R and mass M, with respect to the z-axis that passes through the disk center and is orthogonal to the disk plane.

Solution

The disk can divided into thin rings with infinitesimal mass $dm = \sigma 2\pi r dr$, where $\sigma = \frac{M}{\pi R^2}$ is the superficial mass density. Thus

$$I_z = \int r^2 dm = \int \sigma 2\pi r^3 dr = \frac{\pi \sigma}{2} R^4 = \frac{1}{2} MR^2$$

(E7.3) Determine the momentum of inertia of a homogeneous solid sphere, with radius R and mass M, with respect to the z-axis that passes through the ring center.

Solution

The volumetric mass density of the sphere is

$$\rho = \frac{M}{V} = \frac{M}{\frac{4}{3}\pi R^3}$$

The sphere can be divided into thin disks between $-R \leq z \leq R$ (with z altitude in the sphere). Each disk has infinitesimal mass $= \rho \pi r^2 dz$, where r is the radius of the disk and $r^2 = R^2 - z^2$.

The infinitesimal momentum of inertia of each disk is

$$dI_z = \frac{1}{2} dm r^2 = \frac{1}{2} \rho \pi r^4 dz = \frac{1}{2} \rho \pi \left(R^2 - z^2\right)^2 dz$$

Thus

$$I_z = \tfrac{1}{2}\rho\pi \int_{-R}^{R} (R^2 - z^2)^2 dz$$

$$= \tfrac{1}{2}\rho\pi \int_{-R}^{R} (R^4 - 2R^2 z^2 + z^4) dz$$

$$= \tfrac{1}{2}\rho\pi \left[R^4 z - \tfrac{2R^2 z^3}{3} + \tfrac{z^5}{5} \right]_{-R}^{R}$$

$$= \tfrac{1}{2}\rho\pi \left[\left(R^5 - \tfrac{2R^5}{3} + \tfrac{R^5}{5}\right) - \left(-R^5 + \tfrac{2R^5}{3} - \tfrac{R^5}{5}\right) \right]$$

$$= \tfrac{1}{2}\rho\pi \tfrac{15-10+3+15-10+3}{15} R^5 = \tfrac{1}{2}\rho\pi R^5 \tfrac{16}{15} = \tfrac{8}{15}\rho\pi R^5$$

With the volumetric mass density it can be obtained

$$I_z = \frac{2}{5} M R^2$$

(E7.4) A pulley can be represented as a hollow cylinder with an inextensible wire of negligible mass around it. The cylinder weighs 10 kg and has a radius of 20 cm. The wire is pulled by a small motor that delivers a power of 100 W for 5 s. Determine the final speed of the wire.

Solution

From the power and the time, it is possible to determine the work:

$$W = P/t = 100W/5s = 20J$$

Conservation of mechanical energy can be used:

$$W = E_k \rightarrow W = (1/2) I_z \omega^2$$

Thus $\omega = \sqrt{2W/I_z}$
Considering the momentum of inertia $I_z = mr^2$

$$\omega = \sqrt{2W/mr^2} = \sqrt{\frac{220J}{10kg(0.2m)^2}} = 100 \, rad/s$$

Thus, the velocity is $v = \omega r = 100 \, rad/s \, 0.2 \, m = 20 \, m/s$.

(E7.5) A sphere rotates with angular velocity $\omega_i = 50$ rad/s. The radius of the sphere is $r = 12$ cm. The sphere is dropped on a horizontal plane and the dynamic friction coefficient at the interface $\mu_d = 0.05$. Assuming that the sphere does not bounce on the plane, determine the time required for the sphere to roll without slipping or crawling. In addition, determine how much space the sphere travels in the 2 s following the onset of its pure rolling.

Solution

Considering the mass of the sphere m and its acceleration a, Euler's first law for the sphere is

$$ma = F_f$$

where $F_f = \mu_d R_N = \mu_d mg$ (dynamic friction). Considering the momentum of inertia around the axis z, I_z, and the angular acceleration α, Euler's second law is

$$I_z \alpha = -F_f r$$

Euler's first law can be rewritten as

$$m(dv_{cm}/dt) = \mu_s mg$$

Thus

$$dv_{cm} = \mu_s g \, dt$$

Integrating in t (the initial center of mass velocity is zero)

$$\int dv_{cm} = \int \mu_s g \, dt \rightarrow v_{cm,t} = \mu_s g t$$

For a sphere $I_z = (2/5)mr^2$. Euler's second law can be rewritten

$$(2/5)mr^2 (d\omega/dt) = -\mu_s mgr$$

Integrating in t (the initial angular velocity is ω_i)

$$\int d\omega = -\int (5\mu_s g/2r) dt \rightarrow \omega = \omega_i - (5\mu_s g/2r)t$$

The pure rolling condition is $v_{cm,t_{pr}} = \omega r$, thus

$$\mu_s g t_{pr} = \omega_i r - (5\mu_s g/2) t_{pr}$$

So

$$t_{pr} = 2\omega_i r / 7\mu_s g = \frac{250 \, rad/s \, 0.12 \, m}{70.059.8 \, m/s^2} = 3.499 s$$

To find the space of the sphere in the 2 s following the onset of its pure rolling, it is possible to find the velocity in the pure rolling regime:

$$v_{cm,t_{pr}} = \mu_s g t_{pr} = 0.059.8\, m/s^2 3.499\, s = 1.714\, m/s$$

With the speed, it is possible to determine the space:

$$s_{2s} = v_{cm,t_{pr}} t_{2s} = 1.714\, m/s 2\, s = 3.429\, m$$

s_{2s} can be determined also by finding the angular velocity in the pure rolling regime.

(**E7.6**) A rigid body weighs 2 kg, and the moment of inertia with respect to an axis that is 20 cm from its center of mass is $1\, kgm^2$. What is its moment of inertia with respect to the axis passing through the center of mass?

Solution

For Huygens-Steiner (parallel axes) theorem:

$$I = I_{cm} + Md^2$$

Thus

$$I_{cm} = I - Md^2 = 1\, kgm^2 - 2\, kg(0.2\, m)^2 = 0.92\, kgm^2$$

Exercises

(7.1) A hollow sphere moves by pure rolling on an inclined plane starting from height h. Which statement is true?

(A) $mgh = (1/2)I_z\omega^2 + (1/2)mv^2$ with $I_z = (2/5)mr^2$
(B) $mgh = (1/2)I_z\omega^2 + (1/2)mv^2$ with $I_z = (2/3)mr^2$
(C) $mgh = (1/2)I_z\omega^2$ with $I_z = (2/5)mr^2$
(D) $mgh = (1/2)I_z\omega^2$ with $I_z = (2/3)mr^2$
(E) $mgh = (1/2)mv^2$

(7.2) An hollow cylinder and a solid cylinder rotate without creeping along an inclined plane starting from a height h. At the end of the inclined plane:

(A) The hollow cylinder is twice as fast as the solid one
(B) The solid cylinder is twice as fast as the hollow one
(C) The hollow cylinder is 4/3 times faster than the solid one
(D) Their velocity at the end of the inclined plane is the same one
(E) The hollow cylinder is $\sqrt{1.3}$ times faster than the solid one

(7.3) A 10 kg solid cylinder with a radius of 30 cm, under the effect of a force, rotates 18° with an acceleration of 10 rad/s². How much work is produced by the force?

(A) 1.395 J
(B) 2.783 J

(C) 0.965 J
(D) 4.038 J
(E) 0.231 J

(7.4) A balance is composed of two 1 kg masses held together by a rod of negligible mass. The position vector of the first mass is $\vec{r}_1 = 1\hat{u}_x$ and its velocity is $\vec{v}_1 = 1\hat{u}_y$. The position vector of the second mass is $\vec{r}_1 = -1\hat{u}_x$ and its velocity is $\vec{v}_1 = 1\hat{u}_y$. What is the total angular momentum of the balance?

(A) $\vec{L}_{tot} = 1\hat{u}_z$
(B) $\vec{L}_{tot} = -1\hat{u}_z$
(C) $\vec{L}_{tot} = 2\hat{u}_z$
(D) $\vec{L}_{tot} = -2\hat{u}_z$
(E) $\vec{L}_{tot} = 0$

(7.5) A physical pendulum is made by a solid sphere (m = 5 kg; r = 10 cm) attached to a inextensible and massless rope of length d = 1 m. What is the period of the pendulum?

(A) 2.13 s
(B) 0.74 s
(C) 6.88 s
(D) 4.02 s
(E) 1.22 s

Chapter 8
Fluid Mechanics

Fluid statics

Pascal's principle

The selected volume of liquid is enclosed by a prism with a triangular base with sides a, b, and c and dimension on the z-axis (coming out of the sheet) of length L (Fig. 8.1).

Case of static liquids: there is no movement of any part. The forces are orthogonal to the surfaces, because if there were parallel forces, being able to flow the parts of liquid with respect to each other, there would have motion of the liquid, which is not possible in the static case.

$$p_a = p_b = p_c \tag{8.1}$$

So the pressure does not depend on how the surface of the volume element is inclined, that is, the *pressure propagates in all directions with the same intensity*. The latter is the statement of Pascal's principle.

Stevin's law

Above the liquid there is a pressure p_0 given by atmospheric pressure.

$$p_h = p_0 + \rho g h \tag{8.2}$$

This relationship is called Stevin's law and shows that *in a liquid* with constant density ρ *the pressure increases linearly with depth* (Fig. 8.2).

Archimedes' principle

A body immersed wholly or partially in a fluid receives a downward thrust equal to the weight of the displaced liquid (Fig. 8.3).

Fig. 8.1 Sketch of a triangular prism in a liquid

Fluid statics
Pascal's principle

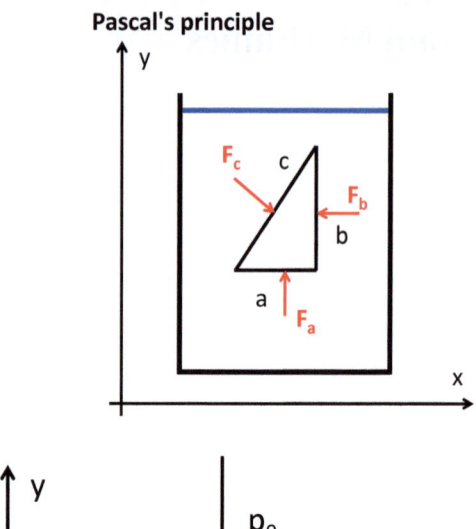

Fig. 8.2 Sketch of a reservoir with liquid with the parameters as mentioned in Eq. 8.2

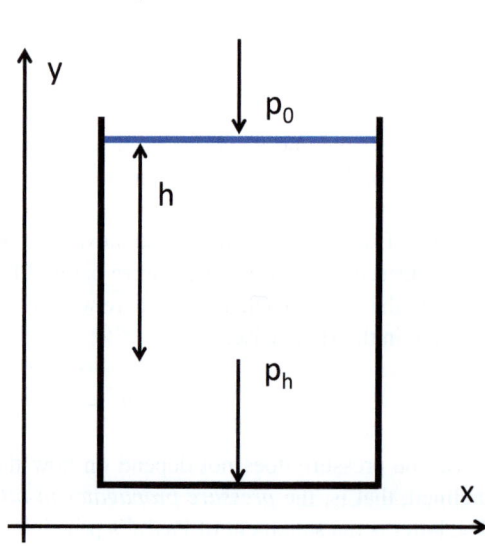

Fig. 8.3 Sketch of a body in a liquid following Archimedes' principle

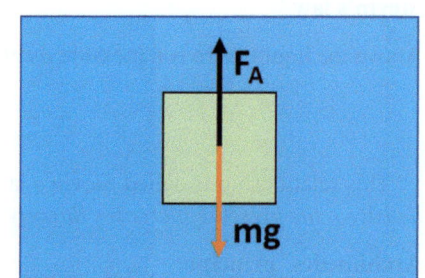

8 Fluid Mechanics

An object with mass m and density ρ_s (s stands for solid) is considered. For the second law of dynamics:

$$mg - F_{Archimedes} = ma \tag{8.3}$$

The solid is removed and the liquid is put back in that volume (with density ρ_L). The force applied to the center of gravity of the water cube is equal to the weight of the liquid (strength of Archimedes):

$$F_{Archimedes} = -\rho_L g V \tag{8.4}$$

Important is the application in the naval field, for ships and submarines. Natural case: swim bladder of fish.

Fluid dynamics

An ideal flow is considered, i.e., incompressible and without friction between molecules. A flow line indicates the path followed by a fluid volume element. In laminar flow, fluid layers flow over each other. In turbulent flow, there is continuous reconfiguration of flow lines (chaotic motion).

The continuity equation for velocities v_1 and v_2 and for tube sections S_1 and S_2 is

$$v_1 S_1 = v_2 S_2 \tag{8.5}$$

The volumetric flow rate is the volume of fluid that flows in a tube over time, and is equal to the product S*v:

$$dV/dt = Sv \tag{8.6}$$

The Bernoulli equation gives a relation between pressure, velocity, and altitude between two points of the tube, with laminar and stationary (velocity depends on space and does not depend on time) flow:

$$p_1 + \rho g h_1 + (1/2)\rho v_1^2 = p_2 + \rho g h_2 + (1/2)\rho v_2^2 \tag{8.7}$$

(please see example E8.4 for a simple application of the Bernoulli equation).

Examples

E8.1) A body is immersed in water at a depth of 20.61 m. What is the pressure exerted on the body?

Solution

Considering the atmospheric pressure $p_0 = 1$ atm $= 1.01 \times 10^5$ Pa $= 1.01 \times 10^5$ N/m^2

$$p = p_0 + \rho g h$$
$$= 1.01 \times 10^5 \tfrac{N}{m^2} + \left(1000 \tfrac{kg}{m^3}\right)(9.8 \tfrac{m}{s^2})(20.61m) \cong 3.03 \times 10^5 \tfrac{N}{m^2}$$

The result is three times the atmospheric pressure, which means that pressure applied by a column of about 10.3 m of water has the same intensity as the atmospheric pressure.

E8.2) With wood of density $\rho_w = 0.5$ kg/L a square raft is constructed. The side dimension of the raft is $L = 2$ m, while its thickness is $th = 15$ cm. How many $m = 80$ kg people can stand on the raft without getting wet?

Solution

The weight of each person is $W = mg = 80 kg\ 9.8\ m/s^2 = 784\ N$.

The Archimedes force for the raft (with the upper surface at the level of the water) is

$$F_A = \rho_{H_2O} V_{raft} g = \rho_{H_2O}(2m 2m 0.15m)g = 5880\ N$$

The total weight of the raft (considering the density of the wood with respect to the one of water) with N persons on top is

$$W_{tot} = 0.5 \rho_{H_2O} V_{raft} g + NW$$

At the equilibrium $F_A = W_{tot}$, which means

$$\rho_{H_2O} V_{raft} g = 0.5 \rho_{H_2O} V_{raft} g + NW$$

To determine N

$$N = \frac{\rho_{H_2O} V_{raft} g - 0.5 \rho_{H_2O} V_{raft} g}{W} = 3.75$$

This means that 3 persons can stand on the raft.

E8.3) A hydraulic press consists of two communicating cylinders of different cross sections, $s_1 = 0.2\ m^2$ and $s_2 = 1\ m^2$, respectively, where two pistons slide. Inside the cylinders is a liquid. If a force $F_1 = 10\ N$ is applied on the smaller piston, what is the force F_2 on the larger piston?

Solution

The relation between the force F_1 and the pressure is

$$p_1 = F_1/S_1$$

For Pascal's principle

8 Fluid Mechanics

$$p_1 = F_1/S_1 = p_2 = F_2/S_2$$

Thus

$$F_2 = \frac{S_2}{S_1} F_1 = \frac{1\ m^2}{0.2\ m^2} 10\ N = 50\ N$$

E8.4) In a building, water enters through a pipeline with a diameter of 4 cm and an absolute pressure of $5 \times 10^5\ Pa$. A 2-cm-diameter pipe carries water to the upper floor, 4 m above the main pipe. If the flow velocity in the main pipe is 1 m/s, find the velocity and pressure at the upper floor.

Solution

To find the velocity:

$$v_1 S_1 = v_2 S_2 \rightarrow v_2 = \frac{S_1}{S_2} v_1 = \frac{(0.02m)^2 \pi}{(0.01m)^2 \pi} 1m/s = 4m/s$$

To find the pressure in the upper floor, the Bernoulli equation can be used:

$$p_1 + \rho g h_1 + (1/2)\rho v_1^2 = p_2 + \rho g h_2 + (1/2)\rho v_2^2$$

Thus

$$\begin{aligned} p_2 &= p_1 - \rho g(h_2 - h_1) - (1/2)\rho(v_2^2 - v_1^2) \\ &= 5 \times 10^5\ Pa - (1 \times 10^3 kg/m^3)(9.8m/s^2)(4m) \\ &\quad -(1/2)(1 \times 10^3 kg/m^3)(16m^2/s^2 - 1m^2/s^2) \\ &= 4.53 \times 10^5\ Pa \end{aligned}$$

Exercises

(8.1) For Stevin's law the pressure at a certain depth in the liquid:

(A) increases linearly with the density of the liquid and decreases linearly with the depth
(B) increases linearly with the density of the liquid and increases linearly with the depth
(C) is the same as the atmospheric pressure
(D) is always smaller than the atmospheric pressure
(E) depends on the density of the body immersed in the liquid

(8.2) A pool is filled with water to a height of 4 m. Calculate the pressure at the bottom of the pool.

(A) 120,800 Pa
(B) 230,120 Pa

(C) 140,200 Pa
(D) 101,000 Pa
(E) 501,000 Pa

(8.3) In a U-shaped open tube, two immiscible liquids of density $\rho_1 = 1$ kg/L and density $\rho_2 = 2$ kg/L, respectively, are at the two ends of the U. What is the height of the first liquid if the first liquid is at the height of 1 m?

(A) 2 m
(B) 1 m
(C) 0.5 m
(D) 4 m
(E) 0.25 m

(8.4) A parallelepiped-shaped object is 0.1 m high and 50% of its volume is immersed in water. Its density is

(A) the same as that of water
(B) $\rho_{object} = (V_{object}/V_{immersed})\rho_{H_2O} = 2\rho_{H_2O}$
(C) $\rho_{object} = (V_{object}^2/V_{immersed})\rho_{H_2O}$
(D) $\rho_{object} = (V_{immersed}/V_{object})\rho_{H_2O} = 0.5\rho_{H_2O}$
(E) $\rho_{object} = \sqrt{(V_{object}/V_{immersed})\rho_{H_2O}} = \sqrt{2\rho_{H_2O}}$

(8.5) A scroll pump is used to bring the volume inside a cylindrical container with a radius of 10 cm to a pressure of 10 Pa. The force exerted from outside on a circular wall of the cylinder is

(A) 3.17 kN
(B) 2.12 kN
(C) 9.91 kN
(D) 10 N
(E) 0 N

Chapter 9
Thermodynamics

Thermodynamics studies the transformations, in terms of mass and energy, of a system.

System: gas, liquid, set of material blocks; environment: what is outside the system; System + environment = thermodynamic universe.

System: (i) open (energy and matter exchange); (ii) closed (no exchange of matter); (iii) isolated (no exchanges).

Thermal equilibrium: two bodies in contact arrive at the same temperature (T).

Thermal expansion:

$$\frac{l(T) - l_0}{l_0} = \lambda(T - T_0) \tag{9.1}$$

$$\frac{V(T) - V_0}{V_0} = \alpha(T - T_0) \tag{9.2}$$

(λ and α phenomenological coefficients)

Temperature scales: Celsius (based on water phase transf.); Kelvin (triple water point).

Zero principle of thermodynamics: two bodies not in contact with T_A and T_B; object with temperature T_C is used to make two measurements. If $T_A = T_C$ and $T_B = T_C$, then $T_A = T_B$ (the two non-contact bodies are at thermal equilibrium).

Food for thought: Thermal expansion is therefore an effective phenomenon for measuring temperature. Other physical phenomena can also be exploited to construct a thermometer. One for example is the Seebeck effect, where a junction of two different materials, e.g., copper-constantan (constantan is a copper-nickel alloy), responds to a change in temperature with a change in electrical voltage.

The opposite of the Seebeck effect is the Peltier effect, where the junction responds to a change in electrical voltage with a change in temperature. With two junctions of the two materials (arranged forming a ring) there is a hot spot and a cold spot. The cold spot can be used to make small cooling systems, even microscopic ones.

Other types of thermometers are based on the detection of infrared radiation, which is proportional to the temperature of objects emitting such radiation.

Observations for ideal gases (sufficiently rarefied with non-interacting molecules):

(i) Boyle-Mariotte (n and T are constant):

$$pV = cost._1 \tag{9.3}$$

(ii) Volta-Guy Lussac (n is constant): isobar:

$$V = V_0(1 + \alpha t) = V_0 \alpha T \tag{9.4}$$

Isochoric:

$$p = p_0(1 + \beta t) = p_0 \alpha T \tag{9.5}$$

(t in Celsius, T in Kelvin, α coeff. dilat. term., β parameter);

(iii) Avogadro (p and T are constant):

$$\frac{V}{n} = cost._2 \tag{9.6}$$

Combining the three observations gives the *state equation of the ideal gases*:

$$pV = nRT = Nk_B T \tag{9.7}$$

(R = 8.314 J/Kmol; $k_B = R/A_0 = 1,3807 \times 10^{-23} J/K$).

Ideal gas transformations: (i) isotherm: $pV = costante$ (equilateral hyperbola on the pV—Clapeyron plane); (ii) isochorous: $p/T = costante$ (line parallel to y-axis on the pV plane); (iii) isobar: $V/T = costante$ (line parallel to x-axis on plane pV). An example of the pV-Clapeyron plane is shown in the Exercise 9.1.

Thermodynamic Work:

$$L = \int \vec{F} \cdot d\vec{r} = pS \int dh = p\Delta V \tag{9.8}$$

(dh: displacement of the piston of gas container). Positive work: system performs work on the environment; Negative work: environment performs work on the system.

Heat:

A material is placed on an ideal tank, able to keep its temperature T_S.

9 Thermodynamics

The material is at temperature T_A. After a time interval $T_A \to T_S$.

Something passes from one body to another with the effect of changing the temperature. It is heat Q:

$$Q = C \Delta T \tag{9.9}$$

The unit of dimension of heat is the calorie (1 cal = 4.186 J).

By mass of material: $C = mc$ (c specific heat).

For moles of ideal gas: $C = nc^*$ (volume cost. $c_V^* = \frac{3}{2}R$ monatomic gas; $c_V^* = \frac{5}{2}R$ diatomic gas; Meyer's relation: $c_p^* = c_V^* + R$).

Joule experiment for the heat (performed in 1850):

The experiment is performed in a adiabatic system, i.e., a system that cannot exchange heat with outside.

Water is inside a closed volume (calorimeter) with a propeller to heat it and a thermometer to measure temperature. The propeller is driven by a mass that can move downward due to gravity on Earth.

Joule sees a change in temperature. So he sees that a mechanical action has the same consequences on the system as an ideal tank.

In the case of water, 1 g of water needs one calorie (cal, unit of heat) to go 14.5 °C to 15.5 °C. Joule sees that 1 cal = 4.186 J. A relationship is found between a unit of heat and a unit of work.

This tells us that heat is a particular form of energy exchange.

Phase transformations: $Q = \lambda m$ (λ latent heat of the transformation and corresponding to a given T).

Thermodynamic cycle: The amount of heat exchanged (in all cycle processes) is equal to work (in all cycle processes): $Q = L$.

First Law of Thermodynamics:

$$Q - L = \Delta U = U_{finale} - U_{iniziale} \tag{9.10}$$

Internal energy depends only on temperature:

$$U = U(T) + U_0 \to \Delta U = Q = nc_V^* \Delta T \tag{9.11}$$

Adiabatic transformations (Q = 0):

$$L = -\Delta U$$
$$TV^{\gamma-1} = const.$$
$$pV^\gamma = const._2$$
$$Tp^{(1-\gamma)/\gamma} = const._3 \tag{9.12}$$

Proof

$$\Delta U + L = nc_V^* \Delta T + p\Delta V \tag{9.13}$$

$$nc_V^* \Delta T + \frac{nRT}{V}\Delta V = 0 \tag{9.14}$$

$$\frac{\Delta V}{V}\frac{nR}{nc_V^*} = -\frac{\Delta T}{T} \tag{9.15}$$

With Mayer's relation:

$$\frac{c_p^* - c_V^*}{c_V^*}\frac{\Delta V}{V} = -\frac{\Delta T}{T} \tag{9.16}$$

Introducing the parameter $= \frac{c_p^*}{c_V^*}$, the expression becomes

$$(\gamma - 1)\frac{\Delta V}{V} = -\frac{\Delta T}{T} \tag{9.17}$$

Integrating the adiabatic transformation (in dV and dT) from A to B:

$$(\gamma - 1)ln\frac{V_B}{V_A} = ln\frac{T_A}{T_B} \tag{9.18}$$

$$ln\left(\frac{V_B}{V_A}\right)^{\gamma-1} = ln\frac{T_A}{T_B} \tag{9.19}$$

which is an equality between logarithmic functions. Thus

$$T_A V_A^{\gamma-1} = T_B V_B^{\gamma-1} \tag{9.20}$$

In general, for an adiabatic transformation between two points it is possible to write

(1) relation between T and V:

$$TV^{\gamma-1} = constant \tag{9.21}$$

and analogously (using the ideal gas law):

(2) relation between p and V:

$$T = \frac{pV}{nR} \tag{9.22}$$

9 Thermodynamics

thus

$$\frac{pV}{nR} V^{\gamma-1} = constant \tag{9.23}$$

$$pV(V^{\gamma-1}) = constant_2 \tag{9.24}$$

so

$$pV^{\gamma} = constant_2 \tag{9.25}$$

(3) relation between p and T:

$$V = \frac{nRT}{p}$$

$$pV^{\gamma} = constant_2 \tag{9.26}$$

$$p\left(\frac{nRT}{p}\right)^{\gamma} = constant_2 \tag{9.27}$$

$$p\left(\frac{T}{p}\right)^{\gamma} = constant_3 \tag{9.28}$$

$$p^{1-\gamma}T^{\gamma} = constant_3 \tag{9.29}$$

$$\sqrt[\gamma]{p^{1-\gamma}}\sqrt[\gamma]{T^{\gamma}} = constant_3 \tag{9.30}$$

$$Tp^{(1-\gamma)/\gamma} = constant_3 \tag{9.31}$$

Thermal machine: thermodynamic system that cyclically passes through the same thermodynamic states. Overall work: area inscribed by the cycle. There is an absorbed heat Q_{Ass} and a heat released Q_{ced}.

Thermal machine efficiency:

$$\eta = L/Q_{Ass} = 1 - |Q_{Ced}|/Q_{Ass} < 1 \tag{9.32}$$

(clockwise cycle).

Efficiency (quality factor) for a refrigeration machine:

$$\varepsilon = COP = f = Q_{Ass}/L \tag{9.33}$$

(counterclockwise cycle).

Carnot machine: 2 isotherms + 2 adiabatic (all reversible).

A *reversible* transformation satisfies two conditions: (1) it operates through states of equilibrium: (i) mechanical (balance between forces and moments); (ii) chemical (absence of chemical reactions or transfers of components); (iii) thermal (same temperature at all times). (2) absence of dissipative forces.

Carnot machine efficiency:

$$\eta = 1 - T_1/T_2 \qquad (9.34)$$

(see example E9.2 for the derivation of the formula).

Second Law of Thermodynamics

Kelvin-Planck statement: it is impossible to carry out a process that has as its sole result the transformation into work of the heat supplied by a source at uniform temperature.

Clausius statement: it is impossible to carry out a process that has as its sole result the transfer of heat from one body to another at a higher temperature.

N.B: the statements are all equivalent (the falsity of one implies the falsity of all the others).

Carnot's theorem: It is impossible for a machine operating between two temperatures to have a higher yield than the reversible machine working between these two temperatures (i.e., a Carnot machine).

Clausius' theorem: On a reversible closed path:

$$\oint \frac{dQ}{T}\bigg|_{rev} = \int_A^B \frac{dQ}{T} + \int_B^A \frac{dQ}{T} = 0 \qquad (9.35)$$

On an irreversible closed path: $\oint \frac{dQ}{T}\big|_{irr} < 0$

The Clausius integral $\int \frac{dQ}{T}$ between two states, by reversible transformation, is equal for every path and depends only on the initial state and final state: $\int_A^B \frac{dQ}{T} = S_B - S_A = \Delta S$ (*entropy change*, state function).

Isothermal transformation	$S_B - S_A = nR\ln\frac{V_B}{V_A} = -nR\ln\frac{P_B}{P_A}$
Isochoric transformation	$S_B - S_A = nc_V\ln\frac{T_B}{T_A} = nc_V\ln\frac{P_B}{P_A}$
Isobar transformation	$S_B - S_A = nc_p\ln\frac{T_B}{T_A} = nc_p\ln\frac{V_B}{V_A}$

9 Thermodynamics

Cycle with an irreversible transformation:

$$\int_A^B \frac{dQ}{T}\bigg|_{rev} = S_B - S_A > \int_{A a B} \frac{dQ}{T}\bigg|_{irr} \tag{9.36}$$

The integral $\int_A^B \frac{dQ}{T}$ gives entropy change only for a reversible transformation. For a transf. irreversible the result of the Clausius integral is not ΔS, but a minor value. To calculate $S_B - S_A$ for such a transf. irreversible $\int_A^B \frac{dQ}{T}$ is calculated by considering a generic reversible transformation between A and B.

Isolated system: $S_B - S_A \geq 0 \Rightarrow S_B \geq S_A$ (entropy *cannot decrease*, constant only for transf. rev.)

$\Delta S \geq 0$ for an isolated system (mathematical formulation of the second law of thermodynamics).

$\Delta S_{Universe} \geq 0 con \Delta S_{Universe} = \Delta S_{System} + \Delta S_{Environment}$ [natural processes are irreversible, so every natural process necessarily takes place in the direction that determines an increase in overall entropy of system + environment (arrow of time)].

Degraded energy:

In an irreversible free expansion

$$L|_{irr} = \int_{V_A}^{V_B} p_{ext} dV = 0 \tag{9.37}$$

In a reversible isotherm

$$\Delta S = nR \ln \frac{V_f}{V_i} \tag{9.38}$$

and

$$L|_{rev} = nRT_0 \ln \frac{V_f}{V_i} \tag{9.39}$$

The difference in energy between the two transformations is

$$\Delta E = \Delta L = L|_{rev} - L|_{irr} = T_0 \Delta S \tag{9.40}$$

Kinetic theory of gases:

Starting assumptions:

(1) gas consisting of equal molecules in a continuous and disordered manner.
(2) molecule/molecule and molecule/wall collisions all elastic.
(3) intermolecular forces only during collisions, due to short-range repulsive forces (neglecting attractive forces between molecules).
(4) size of molecules is much smaller than the average distances between them.

For N molecules of monoatomic gas (with mass m) in the volume V, in a reservoir with lateral size a:

- Impulse of the molecule to the wall (on the x-axis): $\vec{I} = 2mv_x\hat{u}_x$
- Time between collisions (molecule travels for the a section twice, round trip): $t = 2a/v_x$
- Number of collisions (frequency) $= \frac{1}{t} = \frac{v_x}{2a}$
- Force along x: $F_x = 2mv_x \frac{v_x}{2a} = \frac{mv_x^2}{a}$
- Force on the wall for all the molecules: $R_x = \frac{m}{a}\sum_{i=1}^{N} v_{x,i}^2$
- Pressure on the wall (with surface $S = a^2$): $p = \frac{R_x}{S} = \frac{m}{a^3}\sum_{i=1}^{N} v_{x,i}^2 = \frac{Nm}{V}\frac{1}{N}\sum_{i=1}^{N} v_{x,i}^2$
- Mean square velocity: $\langle v_x^2 \rangle = \frac{1}{N}\sum_{i=1}^{N} v_{x,i}^2$
- In each direction: $\langle v_x^2 \rangle = \langle v_y^2 \rangle = \langle v_z^2 \rangle = \frac{\langle v^2 \rangle}{3}$

It leads to the Joule-Clausius-Krönig equation:

$$pV = \frac{1}{3}Nm\langle v^2\rangle; \; pV = \frac{2}{3}N\langle E_k\rangle; \; \langle E_k\rangle = \frac{3}{2}k_B T \qquad (9.41)$$

The Joule-Clausius-Krönig equation relates pressure and velocity to the mean square velocity and, consequently, to the mean kinetic energy. This results in a relation between the temperature and the kinetic energy of the molecules.

Principle of equipartition of energy: $\langle E_k \rangle = \frac{1}{2}k_B T$ ($\updownarrow$ degrees of freedom)
$U = N_A \langle E_k \rangle = \frac{1}{2}RT$ (knowing that $c_V = \frac{dU}{dT}$).

Examples

E9.1) A piece of ice of mass $m_i = 35$ g and at a temperature of $T_1 = 250$ K is immersed in $m_w = 60$ g of water at a temperature of $T_2 = 330$ K. If the system is contained in an adiabatic-walled vessel, calculate the equilibrium temperature of the system. What is the equilibrium temperature?

Solution

Data for ice and water:
$c_i = 2051$ J/kg*K
$c_w = 4186.8$ J/kg*K

$\lambda_{fi} = 3.3 \times 10^5$ J/kg
R = 8.314 J/K*mol.

You have to bring the ice to 273.15 K, when it starts to melt (first term), make it melt (second term), bring it to the equilibrium temperature (third term). On the other side of equal, water at 300 K that gives up heat to the ice and then melts.

$$m_i c_i (273.15 - 250) + \lambda_{fi} m_i + m_i c_w (T - 273.15) = m_w c_w (330 - T)$$

$$23.15 m_i c_i + \lambda_{fi} m_i + T m_i c_w - 273.15 m_i c_w = 330 m_w c_w - T m_w c_w$$

$$T = \frac{330 m_w c_w + 273.15 m_i c_w - 23.15 m_i c_i - \lambda_{fi} m_i}{m_w c_w + m_i c_w} = 275.8 K$$

E9.2) Determine the efficiency of a Carnot machine between the temperatures T_1 and T_2 ($T_2 > T_1$), where all the transformations are reversible.

Solution

The cycle is composed by (a) a reversible isotherm expansion at temperature T_2 between states A and B; (b) a reversible adiabatic expansion between states B and C; (c) a reversible isotherm compression at temperature T_1 between states C and D; (d) a reversible adiabatic compression between states D and A.

(a) Reversible adiabatic expansion at temperature T_2

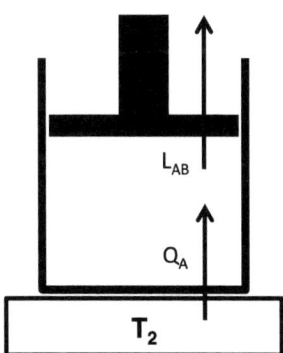

A reversible isotherm is a succession of infinitesimal transformations. For each of these infinitesimal transformations:

- as a result of a decrease in pressure dp there is an expansion of volume dV (mechanical equilibrium), with cooling dT.

- there is a heat transfer dQ from the ideal reservoir at T_2 to the gas and back to temperature T_2 (thermal equilibrium).

For the isotherm $\Delta U = 0 \Rightarrow Q = L$

$$W_{AB} = \int_A^B p\,dV = \int_A^B \frac{nRT_2}{V}dV = nRT_2 \int_A^B \frac{1}{V}dV = nRT_2 \ln\frac{V_B}{V_A}$$
$$\Rightarrow Q_{Assorbito} = Q_A = L_{AB} = nRT_2 \ln\frac{V_B}{V_A}$$

(b) Reversible adiabatic expansion

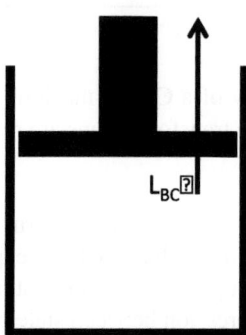

Adiabatic transformation is a transformation isolated from any heat source: Succession of infinitesimal transformations with decrease dp, consequent increase dV with cooling dT.

Between the states B and C, for such adiabatic expansion, there is the relation: $T_2 V_B^{\gamma-1} = T_1 V_C^{\gamma-1}$.

Moreover, $Q = 0 \Rightarrow L = -\Delta U$.

The work done by the gas from B to C is

$$W_{BC} = -\Delta U_{BC} = -nc_v(T_1 - T_2) = nc_v(T_2 - T_1)$$

(c) Reversible isotherm compression at temperature T_1

9 Thermodynamics

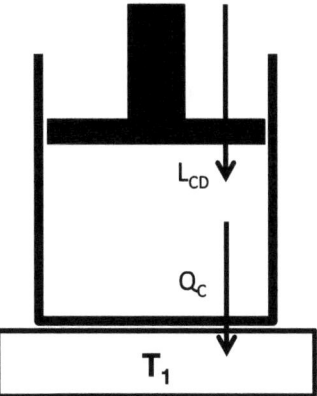

For each of these infinitesimal transformations:

- following an increase in pressure dp there is a decrease in volume dV (mechanical equilibrium), with heating dT.
- there is a heat transfer dQ from the gas to the source at T_1 and back to temperature T_1 (thermal equilibrium).

This heat given up is equal to the work undergone by the gas from C to D:

$$W_{CD} = \int_C^D p \, dV = \int_C^D \frac{nRT_1}{V} dV = nRT_1 \int_C^D \frac{1}{V} dV = nRT_1 \ln \frac{V_D}{V_C}$$
$$\Rightarrow Q_{Ceduto} = Q_C = W_{CD} = nRT_1 \ln \frac{V_D}{V_C}$$

The work is negative since $V_D < V_C$.

(d) Reversible adiabatic compression

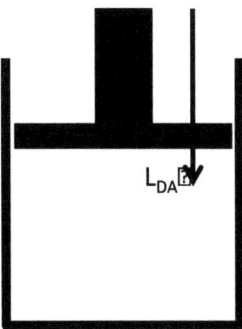

The reversible adiabatic compression is a succession of infinitesimal transformations with increase in dp, subsequent decrease in dV with heating dT.

Between the states D and A, for such adiabatic expansion, there is the relation:
$T_2 V_A^{\gamma-1} = T_1 V_D^{\gamma-1}$
Moreover, $Q = 0 \Rightarrow L = -\Delta U$
The work done on the gas from D to A is

$$L_{DA} = -\Delta U_{DA} = -nc_v(T_2 - T_1) = nc_v(T_1 - T_2) = -L_{BC}$$

Considering all the contributions in the cycle:

$$Q = Q_A + Q_C = W = W_{AB} + W_{BC} + W_{CD} + W_{DA} = W_{AB} + W_{CD}$$

$$\eta = 1 + \frac{Q_C}{Q_A} = 1 + \frac{nRT_1 \ln \frac{V_D}{V_C}}{nRT_2 \ln \frac{V_B}{V_A}} = 1 - \frac{nRT_1 \ln \frac{V_C}{V_D}}{nRT_2 \ln \frac{V_B}{V_A}}$$

Using the following relations for the adiabatic transformations:
$T_2 V_B^{\gamma-1} = T_1 V_C^{\gamma-1}$ and $T_2 V_A^{\gamma-1} = T_1 V_D^{\gamma-1}$
It is possible to write

$$\left(\frac{V_C}{V_D}\right)^{\gamma-1} = \left(\frac{V_B}{V_A}\right)^{\gamma-1} \Rightarrow \frac{V_C}{V_D} = \frac{V_B}{V_A}$$

Thus, the Carnot machine efficiency is

$$\eta = 1 - \frac{T_1}{T_2}$$

The efficiency of the Carnot machine depends only on the temperatures of the isothermal exchanges.

It must be considered that while the two isotherms are processes with heat transfer (where diathermic walls are used), the two adiabatic ones are processes under isolated conditions with respect to any source with heat transfer (adiabatic walls).

E9.3) An amount of heat equal to 300 J is transferred from one ideal reservoir, at a temperature of 150 K, to another ideal reservoir, placed at a temperature of 100 K. Find the entropy variation for the two ideal reservoirs and the universe.

Solution

The entropy variation for the first reservoir is

$$\Delta S_{300\,K} = \frac{300\,J}{150\,K} = 2\,J/K$$

While the entropy variation for the second reservoir is

$$\Delta S_{150\,K} = \frac{300\,J}{100\,K} = 3\,J/K$$

The entropy variation for the universe is given by the difference

$$\Delta S_U = \Delta S_{150K} - \Delta S_{300K} = 3\,J/K - 2J/K = 1\,J/K$$

E9.4) 2 mol of perfect gas at temperature T are contained in a container of volume V. Opening the container expands the gas into a volume ten times larger. What is the entropy variation associated with the free expansion of the gas?

Solution

In this expansion, for heat Q and work W, Q = W. Thus,

$$Q = nRT \int_{V_{in}}^{V_{fin}} \frac{dV}{V} = nRT\ln\frac{10V}{V} = (3mol)RT(\ln 10)$$

Consequently, the entropy variation is given by

$$\Delta S = \frac{Q}{T} = (3\,mol)R(\ln 10)$$
$$= (3\,mol)(8.314\,J/mol\,K)(\ln 10) = 57.43\,J/K$$

Exercises

(9.1) In the thermodynamic cycle (for a mole of monatomic gas) in the figure, the temperature of the BC isotherm is 300 K; moreover, $p_{A,C} = 100$ Pa, $p_B = 200$ Pa, $V_{A,B} = 1$ m³, $V_C = 2$ m³ (R = 8.314 J/K*mol).

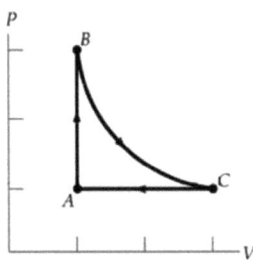

Calculate work and internal energy of the cycle:

(A) L = 618,55 J; $\Delta U = 0$ J
(B) L = 1628,85 J; $\Delta U = 0$ J
(C) L = 2525 J; $\Delta U = 0$ J

(D) L = 1525 J; ΔU = 225,85 J
(E) L = 0 J; ΔU = 685,45 J

(9.2) A lead projectile of mass m = 0.05 kg at temperature t_{Pb} = 20 °C, endowed with velocity v_0 = 100 m/s, sticks horizontally into a block of melting ice of mass M = 0.5 kg, placed on a smooth horizontal plane. Knowing that the specific heat of lead is c_{Pb} = 130 J/kg*°C and the latent heat of fusion of ice λ_{gh} = 3.3106 J/kg, calculate the mass of ice that melted.

(A) 0.3 g
(B) 1.1 g
(C) 2.6 g
(D) 3.7 g
(E) 4.9 g

(9.3) If at constant temperature the number of molecules of ideal gas halves

(A) the product pressure*volume halves
(B) the product pressure*volume doubles
(C) the pressure halves while the volume doubles
(D) the volume halves while the pressure quadruples
(E) volume is necessarily constant so pressure doubles

(9.4) By the second principle of thermodynamics, which process is possible:

(A) a process consisting of only isothermal transformations (for them there is zero energy change)
(B) a process that has as its only result the transformation into work of heat supplied by a source at uniform temperature
(C) in the motion of a body in a plane, the recovery of heat due to sliding friction to create work (and create perpetual motion)
(D) a process where work is produced, but also heat transfer from the higher temperature source to the lower temperature source
(E) a process where the only result is the transfer of heat from one body to another at a higher temperature

9.5) A block of melting ice of mass m_i = 15 g, is placed in a beaker filled with water (m_w = 100 g) at an initial temperature of 20 °C. Calculate the change in total entropy of the system after equilibrium is reached, neglecting heat capacity of the glass and heat exchange with the external environment. (λ_i = 80 cal/g).

(A) How much is the change in total entropy worth?
(B) 0.11 cal/K
(C) 0.88 cal/K
(D) 3.33 cal/K
(E) 0.22 cal/K
(F) 2.05 cal/K

(9.6) What is the equilibrium temperature reached in the exercise 9.5?

   A. 368.87 K
   B. 564.94 K
   C. 125.75 K
   D. 412.46 K
   E. 279.96 K

(9.7) In an adiabatic transformation:

   (A) no work is done
   (B) there is zero change in internal energy
   (C) there is no heat exchange, so $L + \Delta U = 0$
   (D) there is no heat exchange, so $L - \Delta U = 0$
   (E) as in an isochore, the volume does not change

(9.8) According to the second principle of thermodynamics, which process is possible:

   (A) a process consisting of only isothermal transformations (for them there is zero change in energy
   (B) a process that has as its only result the transformation into work of the heat supplied by a source at uniform temperature
   (C) in the motion of a body in a plane, the recovery of heat due to sliding friction to create work (and create perpetual motion)
   (D) a process where work is produced, but also heat transfer from the higher temperature source to the lower temperature source
   (E) a process where the only result is the transfer of heat from one body to another at a higher temperature.

Appendix A

Units of measurement:

- length: meters, m
- time, seconds, s
- speed, m/s
- acceleration, m/s^2
- mass, kg
- momentum, $\frac{kgm}{s}$
- force, $\frac{kgm}{s^2} = N$, Newton
- work and energy, $\frac{kgm^2}{s^2} = Nm = J$, Joule
- power, $\frac{kgm^2}{s^3} = \frac{J}{s} = \frac{Nm}{s} = W$, Watt
- angular momentum $\frac{kgm^2}{s}$
- torque, $m\frac{kgm}{s^2} = mN$ (does not have the same physical meaning as work or energy).

Appendix A

Appendix B

Derivatives

$$Df(x) = \frac{df(x)}{dx} = \lim_{\Delta x \to 0} \frac{\Delta f(x)}{\Delta x}$$

$Dk = 0 \ (k \ real \ number)$	$D tanx = \frac{1}{\cos^2 x}$
$Dx = 1$	$D cotx = -\frac{1}{\sin^2 x}$
$Dx^\alpha = \alpha x^{\alpha-1}$	$De^x = e^x$
$D sinx = cosx$	$Da^x = a^x \log_e a$
$D cosx = -sinx$	$D \log_a x = \frac{1}{x \log_e a}$

$$g(x) = kf(x) \qquad \frac{dg(x)}{dx} = k\frac{df(x)}{dx}$$

$$g(x) = f_1(x) + f_2(x) \qquad \frac{dg(x)}{dx} = \frac{df_1(x)}{dx} + \frac{df_2(x)}{dx}$$

$$g(x) = f_1(x) f_2(x) \qquad \frac{dg(x)}{dx} = \frac{df_1(x)}{dx} f_2(x) + f_1(x)\frac{df_2(x)}{dx}$$

$$g(x) = g[f(x)] \qquad \frac{dg(x)}{dx} = \frac{dg}{df}\frac{df}{dx}.$$

© The Editor(s) (if applicable) and The Author(s), under exclusive license
to Springer Nature Switzerland AG 2023
F. Scotognella, *Exercises in Classical Physics—Mechanics and Thermodynamics*,
Undergraduate Texts in Physics, https://doi.org/10.1007/978-3-031-35074-0

Appendix B

Appendix C

Integrals

Between x_1 and x_2, area below the function:

$$Area = \sum_i f(x_i)\Delta x_i \rightarrow Area = \int_{x_1}^{x_2} f(x)dx$$

$\int dx = x + C$	$\int a^x dx = \frac{a^x}{\log_e a} + C$
$\int x^a dx = \frac{x^{a+1}}{a+1} + C$	$\int \sin x\, dx = -\cos x + C$
$\int \frac{1}{x^2} dx = -\frac{1}{x} + C$	$\int \cos x\, dx = \sin x + C$
$\int \frac{1}{x} dx = \log_e |x| + C$	$\int \frac{1}{\cos^2 x} dx = \tan x + C$
$\int e^x dx = e^x + C$	$\int \frac{1}{\sin^2 x} dx = \cot x + C$

$$\int [f_1(x) + f_2(x)]dx = \int f_1(x)dx + \int f_2(x)dx$$

$$\int k f(x)dx = k \int f(x)dx$$

$$\int f(x)D[g(x)]dx = f(x)g(x) - \int D[f(x)]g(x)dx.$$

Appendix D

Vectors

Scalar quantity: real number

Scalar physical quantity: real number + unit of dimension (e.g., length: 1 m)

Vector:

(i) positive scalar, modulus: $|\vec{v}|$
(ii) direction
(iii) versus

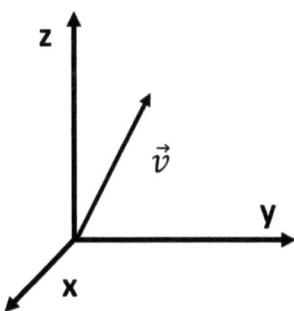

A unit vector is a vector with unitary modulus.

Sum of vectors

With N vectors $\vec{v}_i$, the resulting $\vec{S}$ (sum) vector is given by

$$\begin{cases} S_x = \sum_{i=1}^{N} v_{i,x} \\ S_y = \sum_{i=1}^{N} v_{i,y} \\ S_z = \sum_{i=1}^{N} v_{i,z} \end{cases}$$

Difference of two vectors

With 2 vectors $\vec{v}_1$ and $\vec{v}_2$, the resulting $\vec{D}$ (difference) vector is given by

$$\begin{cases} D_x = v_{1,x} - v_{2,x} \\ D_y = v_{1,y} - v_{2,y} \\ D_z = v_{1,z} - v_{2,z} \end{cases}$$

It is equal to the sum of $\vec{v}_1$ and $-\vec{v}_2$.

Scalar product of two vectors

$$a = |\vec{v}_1||\vec{v}_2|\cos\vartheta$$

ϑ is the angle between the two vectors. A is a scalar.

Vectorial product of two vectors

$$\vec{c} = \vec{a} \times \vec{b} = |\vec{a}||\vec{b}|\hat{n}\sin\beta$$

where β is the angle between $\vec{a}$ and $\vec{b}$, $\hat{n}$ is the unit vector normal to the plane (π in the figure) of $\vec{a}$ and $\vec{b}$.

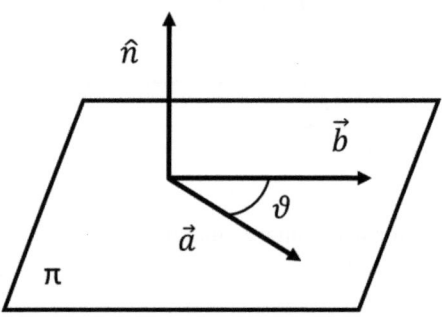

Another way to determine $\vec{c}$

Appendix D

$$\vec{c} = \vec{a} \times \vec{b} = det \begin{bmatrix} \vec{u}_x & \vec{u}_y & \vec{u}_z \\ a_x & a_y & a_z \\ b_x & b_y & b_z \end{bmatrix}$$
$$= (a_y b_z - a_z b_y)\vec{u}_x - (a_x b_z - a_z b_x)\vec{u}_y$$
$$+ (a_x b_y - a_y b_x)\vec{u}_z.$$

Appendix E

Solutions of the problems

1.1. B
1.2. B
1.3. A
1.4. C
1.5. E
1.6. D
1.7. C

2.1. A
2.2. E
2.3. D
2.4. E
2.5. C
2.6. E
2.7. E

3.1. D
3.2. B
3.3. B
3.4. C
3.5. A
3.6. D

4.1. B
4.2. E
4.3. B
4.4. E
4.5. D
4.6. A

5.1. C
5.2. C
5.3. B
5.4. A
5.5. E

6.1. B
6.2. E
6.3. C

7.1. B
7.2. E
7.3. A
7.4. C
7.5. D

8.1. B
8.2. C
8.3. A
8.4. D
8.5. A

9.1. B
9.2. B
9.3. A
9.4. D
9.5. D
9.6. E
9.7. C
9.8. D

Index

A
Acceleration, 2
Adiabatic, 83
Amplitude, 7
Angular acceleration, 5
Angular momentum, 45
Archimedes' principle, 75
Average acceleration, 2
Average velocity, 1

B
Bernoulli equation, 77

C
Calorie, 83
Carnot's theorem, 86
Carnot machine, 86
Center of mass, 36
Center of mass (rigid body), 63
Central force, 46
Centrifugal force, 62
Centripetal force, 45
Circular motion, 5
Clausius statement, 86
Clausius' theorem, 86
Conic function, 47
Conservative force, 26
Continuity equation (for fluids), 77

D
Displacement, 1
Dynamic sliding friction, 17

E
Eccentricity, 43
Elastic constant, 17
Elastic force, 17
Electrostatic force, 49
Entropy, 87
Equilibrium, 7

F
First law of thermodynamics, 83
Fluid, 75
Force, 15
Frequency, 7
Friction, 17

G
Gravitation, 45
Gravitational acceleration, 4
Gravitation constant, 45

H
Harmonic motion, 7
Heat, 36, 82
Hooke's law, 17
Huygens-Steiner (parallel axes) theorem, 64

I
Ideal gas transformations, 82
Impulse, 37
Impulsive, 38
Inclined plane, 21
Inelastic, 36

Inertial frame, 56
Instantaneous acceleration, 2
Instantaneous velocity, 1
Internal energy, 83
Irreversible transformation, 87

J
Joule-Clausius-Krönig equation, 88
Joule experiment, 83

K
Kelvin-Planck statement, 86
Kepler's laws, 43
Kinetic energy, 25
Kinetic theory of gases, 88

L
Laws of dynamics, 15

M
Mass, 16
Mechanical energy, 29

N
Newton, 44
Non-conservative, 28

P
Parabolic motion, 4
Pendulum, 18, 67
Perfectly inelastic collision, 37
Phase, 8
Phase transformation, 83
Potential energy, 26
Potential energy for gravitational interaction, 47
Power, 25
Pressure, 75
Principle of conservation of mechanical energy, 29
Principle of conservation of momentum, 15
Principle of equipartition of energy, 88
Pure rolling, 69

R
Reference frame, 2
Refrigeration machine, 85
Reversible transformation, 86
Rigid body, 35, 63
Rolling, 68
Rotation, 63

S
Scalar product, 25, 104
Second law of thermodynamics, 86
Simple harmonic motion, 7
Spring, 17
State equation of the ideal gases, 82
Static equilibrium, 68
Static sliding friction, 17
Stevin's law, 75

T
Temperature, 81
Tension, 16
Thermal equilibrium, 81
Thermal expansion, 81
Thermal machine, 85
Torque, 45, 67
Trajectory, 2

U
Uniform circular motion, 6
Unit of measurement, 1, 97
Unit vector, 1

V
Vector, 1
Vector product, 6, 104
Viscous friction (Stokes' law), 17
Volumetric flow rate, 77

W
Weight, 16
Work, 25, 82

MIX
Papier aus verantwortungsvollen Quellen
Paper from responsible sources
FSC® C105338

If you have any concerns about our products,
you can contact us on
ProductSafety@springernature.com

In case Publisher is established outside the EU,
the EU authorized representative is:
**Springer Nature Customer Service Center GmbH
Europaplatz 3, 69115 Heidelberg, Germany**

Printed by Libri Plureos GmbH
in Hamburg, Germany